OECD-FAO Agricultural Outlook 2018-2027

This work is published under the responsibility of the Secretary-General of the OECD and the Director-General of the FAO. The opinions expressed and arguments employed herein do not necessarily reflect the official views of OECD member countries, or the governments of the FAO members.

This document, as well as any data and any map included herein are without prejudice to the status of or sovereignty over any territory, to the delimitation of international frontiers and boundaries and to the name of any territory, city or area.

The designations employed and the presentation of material in this information product do not imply the expression of any opinion whatsoever on the part of the Food and Agriculture Organization of the United Nations concerning the legal or development status of any country,territory, city or area or of its authorities, or concerning the delimitation of its frontiers or boundaries.

The names of countries and territories used in this joint publication follow the practice of the FAO.

Please cite this publication as:
OECD/FAO (2018), *OECD-FAO Agricultural Outlook 2018-2027*, OECD Publishing, Paris/Food and Agriculture Organization of the United Nations, Rome.
https://doi.org/10.1787/agr_outlook-2018-en

ISBN 978-92-64-29721-0 (print)
ISBN 978-92-64-06203-0 (PDF)

Series: OECD-FAO Agricultural Outlook
ISSN 1563-0447 (print)
ISSN 1999-1142 (online)

FAO
ISBN 978-92-5-130501-0 (Print and PDF)

The statistical data for Israel are supplied by and under the responsibility of the relevant Israeli authorities. The use of such data by the OECD is without prejudice to the status of the Golan Heights, East Jerusalem and Israeli settlements in the West Bank under the terms of international law.
The position of the United Nations on the question of Jerusalem is contained in General Assembly Resolution 181(II) of 29 November 1947, and subsequent resolutions of the General Assembly and the Security Council concerning this question.

Photo credits: Cover © Original cover concept designed by Juan Luis Salazar. Adaptations by OECD.

Corrigenda to OECD publications may be found on line at: *www.oecd.org/about/publishing/corrigenda.htm*.

© OECD/FAO 2018

You can copy, download or print OECD content for your own use, and you can include excerpts from OECD publications, databases and multimedia products in your own documents, presentations, blogs, websites and teaching materials, provided that suitable acknowledgment of OECD and FAO as source and copyright owner is given. All requests for public or commercial use and translation rights should be submitted to *rights@oecd.org*. Requests for permission to photocopy portions of this material for public or commercial use shall be addressed directly to the Copyright Clearance Center (CCC) at *info@copyright.com* or the Centre francais d'exploitation du droit de copie (CFC) at *contact@cfcopies.com*.

Foreword

The Organisation for Economic Co-operation and Development (OECD) and the Food and Agriculture Organization of the United Nations (FAO) have come together for the 14th year to prepare the *OECD-FAO Agricultural Outlook 2018-2027*. This report is enriched by our close collaboration with contributing member country institutions, specialised commodity bodies, and other partner organisations, and has become an annual benchmark that provides a consistent picture of medium-term trends in global agriculture.

By bringing together evidence-based market and policy information from experts across a wide range of countries, the OECD and FAO are supporting our Members in the pursuit of their shared global priorities. This is particularly the case for the Sustainable Development Goals (SDGs), which aim to end hunger, achieve food security, improve nutrition, and promote sustainable agriculture by 2030. Our joint work on agricultural market projections helps to identify and assess opportunities and threats related to the SDG targets and to the commitments made in the UN Framework Convention on Climate Change's 2015 Paris Agreement. Agriculture not only contributes to climate change (the sector still accounts for more than a fifth of all greenhouse gas emissions), but will also be impacted by climate change. So it is fundamental to promote the adaptation of agricultural sectors through sustainable practices that can also mitigate the impacts of climate change.

Global agro-food trade will also play an increasingly important role in ensuring food security, especially for import-dependent regions. An enabling trade policy environment is a crucial condition to achieve the SDGs and make progress towards zero hunger, particularly in the context of climate change. Building on these efforts, Agriculture Ministers convened at the OECD in 2016 and adopted a Declaration on Better Polices to Achieve a Productive, Sustainable and Resilient Global Food System, which places a high priority on policies that underpin competitive, sustainable, productive and resilient farm and food businesses.

This year's edition of the *Agricultural Outlook* includes a special chapter on the Middle East and North Africa (MENA), a region where conflict and political instability have amplified issues of food insecurity and malnutrition. The need for the region to address these challenges, with limited land and water resources, will be further compounded by the expected impact of more frequent extreme climate-related events. We need to improve the resilience and sustainability of food systems in times of conflict, to valorise resources which are becoming ever more fragile and scarce.

Our partners in the G20 and G7 likewise continue to prioritise food security and agricultural issues on their policy agendas. Along with the *Agricultural Outlook*, the Agricultural Market Information System (AMIS) is part of our wider efforts to provide timely market information to policy makers and global stakeholders. It represents a vital tool that enhances transparency and helps to prevent unexpected price hikes by co-ordinating policy action. AMIS was championed by the G20 and is housed at the FAO with support by numerous international organisations like the OECD.

The challenges we face today cannot be tackled alone. We hope that our collaborative effort on this annual publication will continue to provide governments and all other stakeholders with the evidence base they need to achieve the ambitious and important goals we must meet together.

Angel Gurría	José Graziano da Silva
Secretary-General	Director-General
Organisation for Economic Co-operation and Development	Food and Agriculture Organization of the United Nations

Acknowledgements

The *Agricultural Outlook 2018-2027* is a collaborative effort of the Organisation for Economic Co-operation and Development (OECD) and the Food and Agriculture Organization (FAO) of the United Nations. It brings together the commodity, policy and country expertise of both organisations and input from collaborating member countries to provide an annual assessment of prospects for the coming decade of national, regional and global agricultural commodity markets. The baseline projection is not a forecast about the future, but rather a plausible scenario based on specific assumptions regarding the macroeconomic conditions, the agriculture and trade policy settings, weather conditions, longer term productivity trends and international market developments.

The *Agricultural Outlook* is prepared jointly by the OECD and FAO Secretariats.

At the OECD, the baseline projections and *Outlook* report were prepared by members of the Trade and Agriculture Directorate: Marcel Adenäuer, Jonathan Brooks (Head of Division), Koen Deconinck, Annelies Deuss, Armelle Elasri (publication co-ordinator), Gen Furuhashi, Hubertus Gay (Outlook co-ordinator), Céline Giner, Gaëlle Gouarin, Claude Nenert, Arnaud Pincet and Grégoire Tallard of the Agro-Food Trade and Markets Division, and for fish and seafood by James Innes of the Natural Resources Policy Division. Michael Ryan provided input for the antimicrobial resistance box. The OECD Secretariat is grateful for the contributions provided by visiting experts Joanna Hitchner (United States Department of Agriculture), Roel Jongeneel (Wageningen Economic Research, the Netherlands) and Yu Wen (Chinese Academy of Agricultural Sciences). The partial stochastic modelling builds on work by the Economics of Agriculture Unit of the European Commission's Joint Research Centre, namely Sergio René Araujo Enciso, Simone Pieralli, Thomas Chatzopoulos and Ignacio Pérez Domínguez. The organisation of meetings and publication preparation were provided by Kelsey Burns, Helen Maguire and Michèle Patterson. Technical assistance in the preparation of the *Outlook* database was provided by Eric Espinasse and Frano Ilicic. Many other colleagues in the OECD Secretariat and member country delegations provided useful comments on earlier drafts of the report.

At the Food and Agriculture Organization of the United Nations, the baseline projections and *Outlook* report were prepared by members of the Trade and Markets Division (EST) under the leadership of Boubaker Ben-Belhassen (EST Division Director) and Josef Schmidhuber (EST Division Deputy Director) with the overall guidance of Kostas Stamoulis (Assistant Director-General, Economic and Social Development Department) and by the Economic and Social Development Department Management team. The core projections team consisted of: Katia Covarrubias, Fabio De Cagno, Sergio René Araujo Enciso, Emily Carroll, Gloria Cicerone, Holger Matthey (Team Leader) and Javier Sanchez Alvarez. For fish and seafood, the team consisted of Stefania Vannuccini from the FAO Fisheries and Aquaculture Department, with technical support from Pierre Charlebois. Advice on fishmeal and fish oil issues was provided by Enrico Bachis from the Marine Ingredients Organisation (IFFO). Commodity expertise was provided by

Abdolreza Abbassian, ElMamoun Amrouk, Stanislaw Czaplicki Cabezas, Paulo Augusto Lourenço Dias Nunes, Erica Doro, Alice Fortuna, Jean Luc Mastaki Namegabe, Shirley Mustafa, Adam Prakash, Peter Thoenes, G. A. Upali Wickramasinghe and Di Yang. Input on special topics and boxes was provided by Sabine Altendorf, Tracy Davids, Allan Hruska, Jonathan Pound and Monika Tothova. We thank visiting expert Tracy Davids from the Bureau for Food and Agricultural Policy at the University of Pretoria. Research assistance and database preparation were provided by David Bedford, Julie Claro, Yanyun Li, Emanuele Marocco and Marco Milo. This edition also benefited from comments made by other colleagues from FAO and member country institutions. Araceli Cardenas, Yongdong Fu, Jessica Mathewson, Raffaella Rucci and Juan Luis Salazar provided invaluable assistance with publication and communication issues.

Chapter 2 of the *Outlook*, "The Middle East and North Africa: Prospects and challenges", was prepared by the Secretariats at FAO and OECD. Drafting was led by David Sedik with overall support from the FAO Regional Office for the Near East and North Africa under the leadership of Abdessalam Ould Ahmed, (Assistant Director-General and Regional Representative). Regional projections and analyses were provided by analysts from the Bureau for Food and Agricultural Policy at the University of Pretoria, headed by Prof. Ferdinand Meyer.

Finally, information and feedback provided by the International Cotton Advisory Committee, International Dairy Federation, International Fertilizer Association, International Grains Council, International Sugar Organization, Marine Ingredients Organisation (IFFO) and World Association of Beet and Cane Growers is gratefully acknowledged.

The complete *Agricultural Outlook*, including the fully documented *Outlook* database, including historical data and projections, can be accessed through the OECD-FAO joint internet site: www.agri-outlook.org. The published *Agricultural Outlook 2018-2027* is contained in the OECD's iLibrary.

Table of contents

Tables

Figures

Boxes

Follow OECD Publications on:

 *http://twitter.com/OECD_Pubs*

 *http://www.facebook.com/OECDPublications*

 *http://www.linkedin.com/groups/OECD-Publications-4645871*

 *http://www.youtube.com/oecdilibrary*

 *http://www.oecd.org/oecddirect/*

This book has...

A service that delivers Excel® files from the printed page!

Look for the *StatLinks* at the bottom of the tables or graphs in this book. To download the matching Excel® spreadsheet, just type the link into your Internet browser, starting with the *http://dx.doi.org* prefix, or click on the link from the e-book edition.

Follow FAO on:

Food and Agriculture Organization of the United Nations

 *www.twitter.com/FAOstatistics*
www.twitter.com/FAOnews

 *www.facebook.com/UNFAO*

 *www.linkedin.com/company/fao*

 *www.youtube.com/user/FAOoftheUN*

Acronyms and abbreviations

ACP	African Caribbean and Pacific countries
AI	Avian Influenza
AEU	Additional ethanol use
AMIS	Agricultural Market Information System
AMR	Antimicrobial resistance
ARC	Agricultural Risk Coverage (US Farm Bill Instrument)
ASF	African Swine Fever
bln	Billion
bln L	Billion litres
BRIC	Emerging economies of Brazil, Russian Federation, India and China
BRICS	Emerging economies of Brazil, Russian Federation, India, China and South Africa
BRIICS	Emerging economies of Brazil, Russian Federation, India, Indonesia, China and South Africa
bln t	Billion tonnes
CAP	Common Agricultural Policy (European Union)
CFP	Common Fisheries Policy (European Union)
CETA	Comprehensive Economic and Trade Agreement
ChAFTA	China-Australia Free Trade Agreement
CIF	Cost, insurance and freight
CIS	Commonwealth of Independent States
CPI	Consumer Price Index
CPIF	Consumer Price Index for Food
CPTPP	Trans-Pacific Partnership
CRP	Conservation Reserve Program (United States)
cts/lb	Cents per pound
CVD	Countervailing duty
c.w.e.	Carcass weight equivalent
DDGs	Dried Distiller's Grains
dw	Dry weight
dwt	Dressed weight
EBA	Everything-But-Arms Initiative (European Union)
EISA	Energy Independence and Security Act of 2007 (United States)
El Niño	Climatic condition associated with the temperature of major sea currents
EMEs	Emerging Market Economies
EPA	US Environmental Protection Agency
EPAs	Economic Partnership Agreements
ERS	Economic Research Service of the US Department for Agriculture
ESCWA	United Nations Economic and Social Commission for Western Asia
est	Estimate
EU	European Union
EU15	Fifteen member states that joined the European Union before 2004
EU28	Twenty eight member states of the European Union
FAO	Food and Agriculture Organization of the United Nations
FDP	Fresh dairy products

FDI	Foreign direct investment
FFV	Flex-fuel Vehicles
FOB	Free on board (export price)
FMD	Foot and Mouth Disease
FTA	Free Trade Agreement
G-20	Group of 20 important developed and developing economies (see Glossary)
GCC	Gulf Co-operation Council
GDP	Gross domestic product
GDPD	Gross domestic product deflator
GHG	Greenhouse gas
GIEWS	Global Information and Early Warning System on Food and Agriculture
GM	Genetically modified
GVCs	Global value chains
ha	Hectares
HFCS	High fructose corn syrup
hl	Hectolitre
IEA	International Energy Agency
IFA	International Fertilizer Association
IFAD	International Fund for Agricultural Development
IFPRI	International Food Policy Research Institute
IGC	International Grains Council
ILUC	Indirect Land Use Change
IMF	International Monetary Fund
IPCC	Intergovernmental Panel on Climate Change
IUU	Illegal, unreported and unregulated (fishing)
kg	Kilogrammes
kha	Thousand hectares
kt	Thousand tonnes
La Niña	Climatic condition part of El Niño-Southern Oscillation (see Glossary)
LAC	Latin America and the Caribbean
lb	Pound (weight)
LDCs	Least Developed Countries
LED	Light-emitting diode
lw	Live weight
MBM	Meat and bone meal
MDGs	Millennium Development Goals
MENA	Middle East and North Africa
MERCOSUR	Mercado Común del Sur / Common Market of South America
MFA	Multi-fibre Arrangement
Mha	Million hectares
Mn	Million
Mn L	Million litres
MPS	Market Price Support
Mt	Million tonnes
NAFTA	North American Free Trade Agreement
NCDs	Non-communicable diseases
NRA	Nominal rate of assistance
OECD	Organisation for Economic Co-operation and Development
OIE	World Organisation for Animal Health
OLS	Ordinary Least Squares
OPEC	Organization of Petroleum Exporting Countries
p.a.	Per annum
PCE	Private consumption expenditure

PEDv	Porcine Epidemic Diarrhoea virus
PLC	Price Loss Coverage (US Farm Bill instrument)
PoU	Prevalence of Undernourishment
PPI	Producer Price Index
PPP	Purchasing power parity
PSE	Producer Support Estimate
R&D	Research and development
RED	Renewable Energy Directive in the European Union
RFS / RFS2	Renewable Fuels Standard in the United States, part of the Energy Policy Act
RIN	Renewable Identification Numbers prices
rse	Raw sugar equivalent
RTA	Regional Trade Agreements
r.t.c.	Ready to cook
r.w.e.	Retail weight equivalent
SDG	Sustainable Development Goals
SFP	Single Farm Payment (European Union)
SMP	Skim milk powder
SPS	Single payment scheme (European Union)
SSA	Sub-Saharan Africa
SSR	Self-sufficiency Ratio
t	Tonnes
t/ha	Tonnes/hectare
TFP	Total Factor Productivity
TPP	Trans-Pacific Partnership
tq	Tel quel basis (sugar)
TRQ	Tariff rate quota
UN	The United Nations
UNDP	United Nations Development Programme
UNEP	United Nations Environment Programme
UNFCCC	United Nations Framework Convention on Climate Change
UNICEF	United Nations Children's Fund
URAA	Uruguay Round Agreement on Agriculture
US	United States
USDA	United States Department of Agriculture
VIFEP	Vietnam Institute of Fisheries and Economic Planning
WB	World Bank
WFP	World Food Programme
WHO	World Health Organization
WITS	World Integrated Trade Solution
WMP	Whole milk powder
wse	White sugar equivalent
WTO	World Trade Organization
WWF	World Wide Fund for Nature

Currencies

ARS	Argentinean peso
AUD	Australian dollars
BDT	Bangladeshi taka
BRL	Brazilian real
CAD	Canadian dollar
CLP	Chilean peso
CNY	Chinese yuan renminbi

DZD	Algerian dinar
EGP	Egyptian pound
EUR	Euro (Europe)
IDR	Indonesian rupiah
INR	Indian rupees
JPY	Japanese yen
KRW	Korean won
MXN	Mexican peso
MYR	Malaysian ringgit
NZD	New Zealand dollar
PKR	Pakistani rupee
RUB	Russian ruble
SAR	Saudi riyal
THB	Thai baht
TRL	Turkish lira
UAH	Ukrainian grivna
USD	US dollar
UYU	Uruguayan peso
ZAR	South African rand

Executive Summary

The *Agricultural Outlook 2018-2027* is a collaborative effort of the OECD and FAO prepared with input from the experts of their member governments and from specialist commodity organisations. It provides a consensus assessment of the ten-year prospects for agricultural and fish commodity markets at national, regional and global levels. This year's edition contains a special chapter on the prospects and challenges of agriculture and fisheries in the Middle East and North Africa.

A decade after the food price spikes of 2007-8, conditions on world agricultural markets are very different. Production has grown strongly across commodities, and in 2017 reached record levels for most cereals, meat types, dairy products, and fish, while cereal stock levels climbed to all-time highs. At the same time, demand growth has started to weaken. Much of the impetus to demand over the past decade came from rising per capita incomes in the People's Republic of China (hereafter "China"), which stimulated the country's demand for meat, fish and animal feed. This source of demand growth is decelerating, yet new sources of global demand are not sufficient to maintain overall growth. As a result, prices of agricultural commodities are expected to remain low. Current high stock levels also make a rebound unlikely within the next few years.

The weakening of demand growth is expected to persist over the coming decade. Population will be the main driver of consumption growth for most commodities, even though the rate of population growth is forecast to decline. Moreover, per capita consumption of many commodities is expected to be flat at a global level. This is notable for staple foods such as cereals and roots and tubers, where consumption levels are close to saturation levels in many countries. By contrast, demand growth for meat products is slowing due to regional variation in preferences and disposable income constraints, while demand for animal products such as dairy is set to expand faster in the coming decade.

For cereals and oilseeds, the foremost source of demand growth will be feed, closely followed by food. A large share of additional feed demand will continue to come from China. Feed demand growth is nevertheless projected to slow globally, despite livestock production intensification. Much of the additional food demand will originate in regions with high population growth such as Sub-Saharan Africa, India, and the Middle East and North Africa.

The demand for cereals, vegetable oil and sugar cane as inputs into the production of biofuels is expected to grow much more modestly than in the last decade. Whereas in the past decade the expansion of biofuels led to more than 120 Mt of additional cereals demand, predominately maize, this growth is expected to be essentially zero over the *Outlook* period. In developed countries, existing policies are not likely to support much further expansion. Future demand growth will therefore come predominantly from developing countries, several of which have introduced policies favouring biofuels use.

The exceptions to the broad pattern of slowing per capita demand growth come from sugar and vegetable oils. The per capita intake of sugar and vegetable oil is expected to increase in the developing world, as urbanisation in developing countries leads to a greater demand for processed and convenience foods. Changes in levels of food consumption and the composition of diets imply that the "triple burden" of undernourishment, over-nourishment and malnutrition will persist in developing countries.

Global agricultural and fish production is projected to grow by around 20% over the coming decade, but with considerable variation across regions. Strong growth is expected in Sub-Saharan Africa, South and East Asia, and the Middle East and North Africa. By contrast, production growth in the developed world is expected to be much lower, especially in Western Europe. The growth in production will be achieved primarily from intensification and efficiency gains and partially from an enlargement of the production base through herd expansion and the conversion of pasture to cropland.

With slower consumption and production growth, agriculture and fish trade is projected to grow at about half the rate of the previous decade. Net exports will tend to increase from land abundant countries and regions, notably in the Americas. Countries with high population densities or high population growth, in particular in the Middle East and North Africa, Sub-Saharan Africa and in Asia, will see rising net imports.

For nearly all agricultural products, exports are projected to remain concentrated among stable groups of key supplying countries. A notable change is the emerging presence of the Russian Federation and Ukraine on world cereal markets, which is expected to persist. The high concentration of export markets may increase the susceptibility of world markets to supply shocks, stemming from natural and policy factors.

As a baseline projection, the *Agricultural Outlook 2018-2027* assumes policies currently in place will continue into the future. Beyond the traditional risks that affect agricultural markets, there are increasing uncertainties with respect to agricultural trade policies and concerns about the possibility of rising protectionism globally. Agricultural trade plays an important role in ensuring food security, underscoring the need for an enabling trade policy environment.

Middle East and North Africa

This year's special chapter focuses on the Middle East and North Africa, where rising food demand and limited land and water resources lead to rising import dependence for basic food commodities. Many countries spend a large share of their export earnings on food imports. Food security is threatened by conflict and political instability.

The region's agriculture and fish production is projected to increase about 1.5% p.a., mainly through productivity improvements. Policies in the region support grain production and consumption, with the result that 65% of cropland is planted with water-thirsty cereals, in particular wheat, which accounts for a large share of calorie intake. Diets are projected to remain high in cereals and sugar, with low protein intake from animal sources.

An alternative approach to food security would reorient policies away from supporting cereals towards rural development, poverty reduction and support for production of higher-value horticulture products. Such a change would also contribute to more diversified and healthier diets.

Chapter 1. Overview

This chapter provides an overview of the latest set of quantitative medium-term projections for global and national agricultural markets. The projections cover consumption, production, stocks, trade and prices for 25 agricultural products for the period 2018 to 2027. The weakening of demand growth is expected to persist over the coming decade. Population will be the main driver of consumption growth for most commodities, even though the rate of population growth is forecast to decline. Per capita consumption of many commodities is expected to be flat at a global level. Consequently, the slower growing demand for agricultural commodities is projected to be matched by efficiency gains in production which will keep real agricultural prices relatively flat. Beyond the traditional risks that affect agricultural markets, there are increasing uncertainties with respect to agricultural trade policies and concerns about the possibility of rising protectionism globally

Introduction

The *Agricultural Outlook* presents a baseline scenario for the evolution of agricultural and fish commodity markets at national, regional and global levels over the coming decade (2018-2027). The projections rely on input from country and commodity experts and from the OECD-FAO Aglink-Cosimo model of global agricultural markets. This economic model is also used to ensure the consistency of baseline projections.

The projections are influenced both by current market conditions and by assumptions on the macro-economic, demographic and policy environment. These assumptions are detailed at the end of this chapter (Box 1.6) and in the commodity chapters. The sensitivity of the *Outlook* to these assumptions is discussed later in the chapter.

For the coming decade, economic growth of 1.8% per annum is expected for OECD countries, broadly the same pace as over the last decade (1.7% p.a.). Growth is projected to slow for the People's Republic of China (hereafter "China") but accelerate in India compared with the past decade. Following the strong increase in 2017, nominal oil prices are expected to increase at an average rate of 1.8% per year over the outlook period, from an average price of USD 43.7 per barrel in 2016 to USD 76.1 per barrel by 2027.

The *Outlook* assumes current policy settings continue into the future. In particular, the decision of the United Kingdom to leave the European Union is not included in the projections as the terms of departure have not yet been determined. Projections for the United Kingdom are therefore retained within the European Union aggregate.

Current market conditions for the different commodities included in the *Outlook* are summarised in Figure 1.1, which shows the evolution of production and prices during the base period (2015-17) compared to average levels over the past decade. For most cereals, meat types, dairy products and fish, 2017 production levels exceeded even the high levels recorded last year.

Despite a global economic recovery and higher oil prices, prices for most agricultural commodities did not change much in 2017 compared to the previous year, except for dairy and sugar. Dairy markets were in flux, with low prices in 2016 followed by a recovery in 2017 and a 65% spike in butter prices in the first half of the year which eventually came back down by the end of the year. The recovery of sugar production after two years of shortage contributed to a decline in prices.

These current market conditions form the backdrop for the ten-year projections of consumption, production, trade and prices presented in the next sections.

Figure 1.1. Market conditions for key commodities

Current market conditions

Cereals: World production reached a new record in 2017, as maize and rice output levels surpassed historical levels. Global supplies have exceeded demand for several years, resulting in a significant build-up of inventories and low prices.

Oilseeds: Soybean production declined slightly in the 2017 marketing year, although aggregate production of other oilseeds remained stable. Demand growth for protein meals was lower than in 2016. Overall, there were no major disruptions.

Sugar: After two years of supply shortage, production in 2017 rose and was close to the 2012 record. Sugar prices fell after having increased strongly in 2016. The growth in demand increased in countries with low per capita consumption. Global import demand continued to decline, in part as demand from China waned.

Meat: World meat production increased by 1.2% in 2017. Much of the increase in production originated in the United States, but other main contributors to growth include Argentina, China, India, Mexico, the Russian Federation and Turkey. After declining in 2016, international meat prices increased by 9% in 2017 (as measured by the FAO Meat Price Index), as import demand grew. The highest price increase was noted for sheep meat.

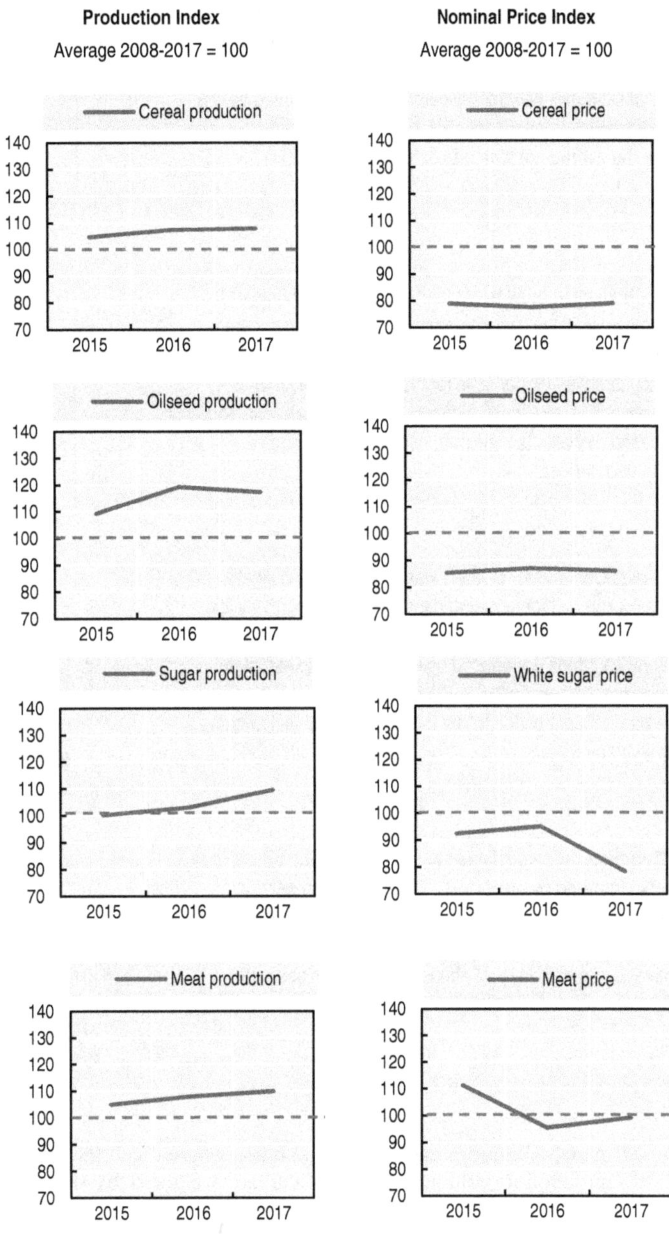

Dairy: Global dairy markets experienced strong price increases in 2017. After an initial increase of 65% in the first half of the year, butter prices came back down by the end of 2017. The price of whole milk powder increased by 46%; the price of skim milk powder, by contrast, only increased by 3%. World production experienced a modest growth of 0.5%, below the average growth rate of the last decade.

Fish: Production grew faster than in 2016, as catches of anchoveta in South America recovered while aquaculture continued to grow at 4% per year. As in recent years, aquaculture was responsible for most of the growth in production. Despite these higher production levels, fish prices increased globally as improving economic conditions stimulated demand.

Biofuels: Demand for biofuels was sustained by obligatory blending and by higher demand for fuel due to relatively low energy prices, despite increasing crude oil prices in 2017. Several countries announced policy decisions to stimulate demand for biofuels in 2017. Prices for ethanol and biodiesel diverged: ethanol prices fell by 2.3% while biodiesel prices increased by 8%.

Cotton: Production continued to recover from the strong drop in 2015, growing by around 9%. Production increased in almost all major cotton producing countries except for China. Despite an increase in world demand, global stocks grew and remained at a high level of almost nine months of world utilisation.

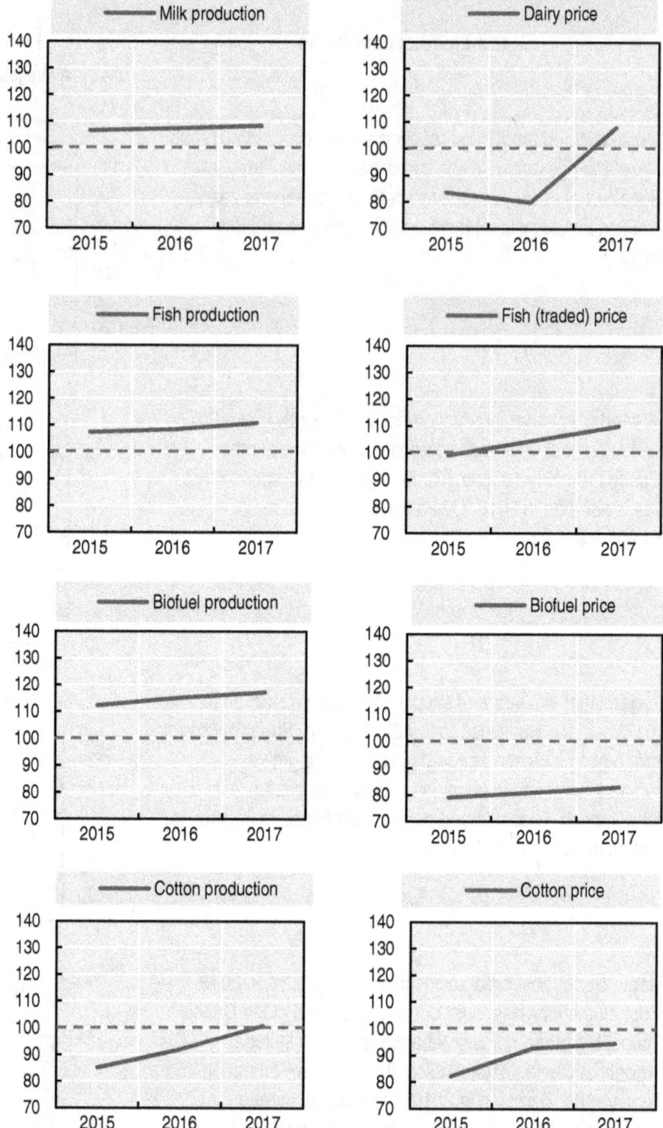

Note: All graphs expressed as an index where the 2008-2017 average is set to 100. Production refers to global production volumes; price indices are weighted by the average global production value in the preceding decade as measured at international prices. More information on market conditions and evolutions by commodity can be found in the commodity snapshot tables in the Annex and the online commodity chapters.
Source: OECD/FAO (2018), "OECD-FAO Agricultural Outlook", OECD Agriculture statistics (database), http://dx.doi.org/10.1787/agr-outl-data-en.

StatLink http://dx.doi.org/10.1787/888933741903

Consumption

Agricultural commodities are consumed mainly as food, feed, and in industrial applications including fuel. Food demand is influenced by population and income growth, and increasingly also by trends in dietary patterns and consumer preferences. Demand for animal feed is closely linked to the human consumption of livestock products, such as meat, eggs and milk, but also by the evolution of livestock production technology. Industrial uses of agricultural commodities (mostly as biofuel and as input in the chemical industry) are shaped by general economic conditions, as well as regulatory policies and technological advances. Moreover, the relative importance of each use varies by commodity, by region, and by level of economic development.

Over the last ten years, agricultural markets experienced a strong increase in demand across a wide range of commodities. Much of that growth was attributable to non-food uses of agricultural commodities, mostly feedstock for biofuel and animal feed. While food demand stagnated in the developed world, biofuel mandates led to increased demand for maize, sugarcane and vegetable oils as feedstock. In parallel, rising incomes in China and other emerging economies raised demand for meat. This in turn drove an intensification of livestock production which boosted demand for animal feed on global markets. Together, these sources of demand growth contributed to real agricultural prices remaining above the levels seen in the early 2000s, fuelling production growth worldwide.

Biofuels and Chinese demand growth will continue to play a role in global agricultural markets. However, their relevance is diminishing and they are not fully being replaced by new sources of demand growth, whether for food, feed, or fuel uses.

In terms of food demand, per capita consumption of many commodities is expected to be flat at a global level. This is not only expected for staple foods such as cereals and roots and tubers, where consumption levels are close to saturation levels in many countries, but also for meat. Some low-income regions which currently have low per capita consumption levels of meat, such as Sub-Saharan Africa, are not expected to increase these levels significantly due to a lack of sufficient income growth. Some emerging economies, in particular China, have already transitioned to relatively high levels of per capita meat consumption. In India, where income growth is stronger, dietary preferences translate rising incomes into an increased per capita demand for dairy as preferred animal protein, rather than meat.

One implication of relatively flat per capita food consumption is that population growth will be the principal determinant of food demand growth, even though global population is projected to grow at a lower rate in the coming decade. The bulk of additional food consumption in the coming decade will originate in regions with high population growth such as Sub-Saharan Africa, India, and the Middle East and North Africa (the focus of Chapter 2). Demand patterns in these regions will increasingly influence international agricultural markets.

The demand for feed, meanwhile, will continue to outpace food demand as livestock production intensifies. A large share of additional feed demand will come from China, as in the previous decade. Yet, compared with the previous decade, demand growth for feed slows down.

Finally, recent developments in biofuel policies combined with the assumption of a relatively moderate increase in the crude oil price suggest a more modest growth in the use of agricultural commodities in the production of biofuels.

As a result of these developments in food, feed, and fuel uses of agricultural commodities, a slower growth in global demand for agricultural commodities is expected in the coming decade (Figure 1.2).

Figure 1.2. Annual growth in demand for key commodity groups, 2008-17 and 2018-27

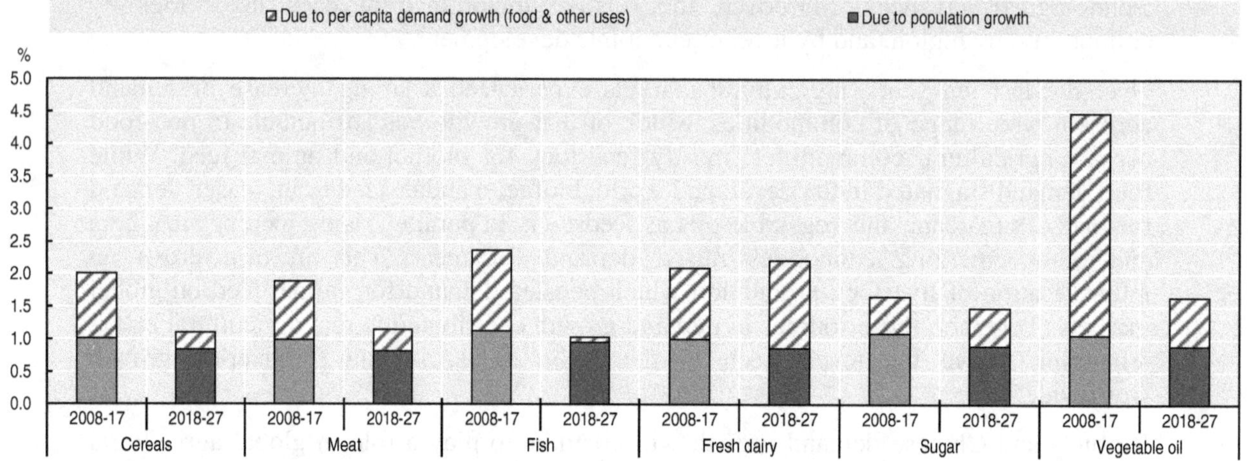

Note: The population growth component is calculated assuming per capita demand remains constant at the level of the year preceding the decade. Growth rates refer to total demand (for food, feed and other uses).
Source: OECD/FAO (2018), "OECD-FAO Agricultural Outlook", OECD Agriculture statistics (database), http://dx.doi.org/10.1787/agr-outl-data-en.

StatLink http://dx.doi.org/10.1787/888933741922

For cereals, meat, fish, and vegetable oil, growth rates are around half their rates in the previous decade. The slowdown is particularly pronounced for vegetable oil, which was the fastest-growing commodity over the past decade, as biofuel policies, industrial uses (for paints, lubricants, detergents, etc.) and a strong growth in food use supported demand. Despite the slowdown, vegetable oil remains one of the fastest growing commodities in the *Outlook*, together with fresh dairy products and sugar.

Food: Population and income growth spurs demand in the developing world

Food consumption will continue to expand due to population growth and higher per capita income for most commodities with the developing world as the source of most demand growth over the coming ten years (Figure 1.3). Sub-Saharan Africa and India will account for a large share of the additional food demand for cereals in the coming decade. Consumption of dairy products and vegetable oil in India will underpin growth in these commodities over the next ten years, while China continues to account for a large share of demand growth for meat and fish.

Figure 1.3. Regional contributions to food demand growth, 2008-17 and 2018-27

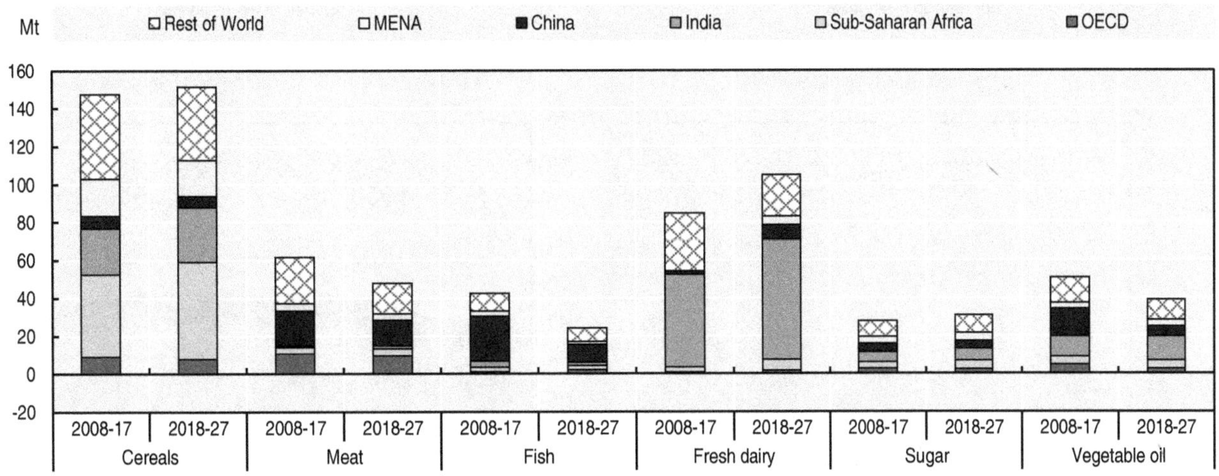

Mt
Rest of World, MENA, China, India, Sub-Saharan Africa, OECD
160, 140, 120, 100, 80, 60, 40, 20, 0, -20
2008-17 | 2018-27 — Cereals
2008-17 | 2018-27 — Meat
2008-17 | 2018-27 — Fish
2008-17 | 2018-27 — Fresh dairy
2008-17 | 2018-27 — Sugar
2008-17 | 2018-27 — Vegetable oil

Note: Each column shows the increase in global demand over a ten-year period, split by region, for food uses only. MENA stands for Middle East and North Africa, and is defined as in Chapter 2.
Source: OECD/FAO (2018), "OECD-FAO Agricultural Outlook", OECD Agriculture statistics (database), http://dx.doi.org/10.1787/agr-outl-data-en.

StatLink http://dx.doi.org/ 10.1787/888933741941

The important contribution from Sub-Saharan Africa and India reflects in large measure continued strong population growth in these regions (Figure 1.4). The global population growth rate is expected to fall from 1.1% at present to 0.9% per year in 2027. Since around 2013, growth has also been falling in absolute terms, although world population will still grow by around 74 million people per year by 2027. Most of this growth occurs in Sub-Saharan Africa and India, as well as the Middle East and North Africa. Population growth in Sub-Saharan Africa is accelerating in absolute terms: while the region's population increased by 27 million in 2017, this rate will increase to 32 million extra people per year in 2027.

In addition to population growth, food demand is influenced by the growth of per capita incomes. The macro-economic assumptions underlying this *Outlook* suggest strong growth in per capita GDP in India (6.3% p.a.) and China (5.9% p.a.). For Sub-Saharan Africa, 2.9% p.a. per capita growth is expected over the coming decade, but with variations across the continent. Moreover, high growth in average incomes does not necessarily translate to income growth for poorer households. Per capita food demand in Sub-Saharan Africa is therefore expected to remain at relatively low levels.

Finally, differences in dietary preferences shape demand patterns. While income growth in China in the last decade led to increased demand for meat and fish, rising incomes in India are mostly expected to lead to higher consumption of dairy products as the preferred source of animal proteins. The interplay of such regional differences in population growth, income growth and dietary preferences thus result in different developments for individual commodities.

Figure 1.4. World population growth, 1998-2027

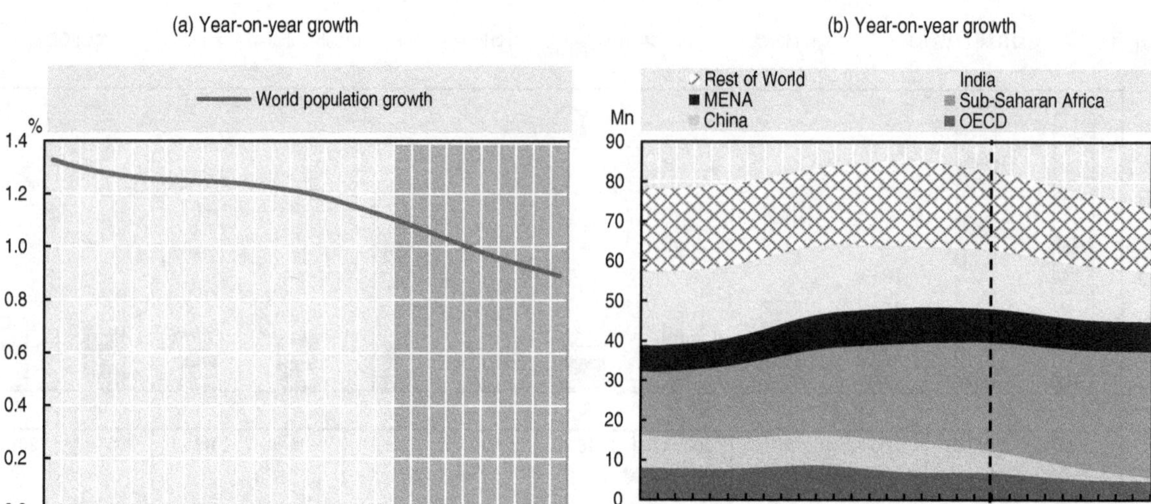

Note: MENA Middle East and North Africa.
Source: OECD/FAO (2018), "OECD-FAO Agricultural Outlook", OECD Agriculture statistics (database), http://dx.doi.org/10.1787/agr-outl-data-en.

StatLink http://dx.doi.org/ 10.1787/888933741960

Cereals: Growth in food consumption determined mainly by population growth

Figure 1.5 shows the level and composition of per capita consumption of cereals in main regions, illustrating the high per capita consumption of cereals around the world, especially in the Middle East and North Africa. It also shows the continued dominance of wheat and rice across regions, except in Sub-Saharan Africa. In this region, white maize plays a major role in cereals consumption, and in calorie intake, as discussed in Box 1.1.

Globally, per capita cereals consumption increases by less than 2% over the coming decade. This slow growth is explained in large part due to the near-saturation level of cereals consumption in many regions across the world. Per capita food consumption of cereals is expected to grow only in low-income regions such as Sub-Saharan Africa, where per capita consumption increases by 6% over the next decade. In such low-income regions, cereals account for about two-thirds of dietary energy, compared to about one-third in developed regions.

Given relatively flat per capita consumption, population growth is the main determinant of growth in the coming decade, and the regions with the greatest population expansion (Sub-Saharan Africa, India, the Middle East and North Africa) will also account for the bulk of the additional food consumption of cereals.

Figure 1.5. Cereals: Availability for food consumption

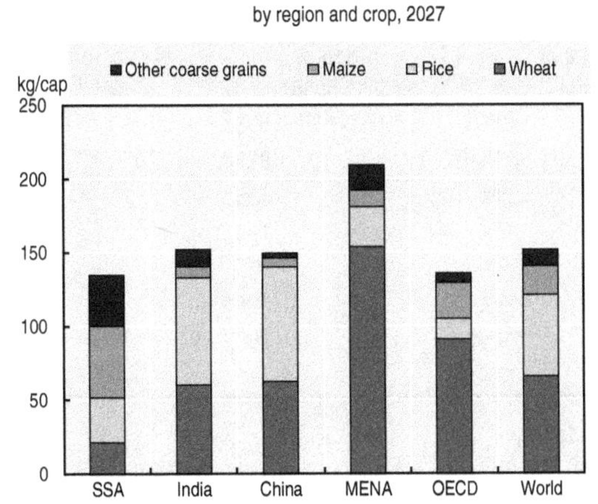

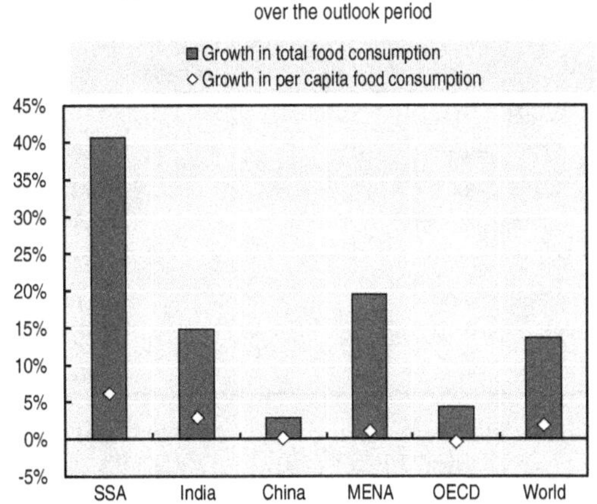

Note: SSA is Sub-Saharan Africa; MENA is Middle East and North Africa. The *Agricultural Outlook* measures consumption in terms of food availability and hence does not account for waste.

Source: OECD/FAO (2018), "OECD-FAO Agricultural Outlook", OECD Agriculture statistics (database), http://dx.doi.org/10.1787/agr-outl-data-en.

StatLink http://dx.doi.org/10.1787/10.1787/888933741979

Box 1.1. White maize and food security in Sub-Saharan Africa

Maize is a primary source of calories in Sub-Saharan Africa[1], contributing about 19% of calorie availability on average (Table 1.1). Consumers prefer non-GMO white maize, typically produced locally or imported from within the region. Production is mostly low-input, rain-fed and smallholder-based, resulting in significant local yield variability. Local deficits are offset mostly through intra-country and regional trade; where such flows are impeded, production volatility threatens local food security.

Regional trade within Sub-Saharan Africa accounts for about 5% of food consumption, but this figure varies considerably by country. South Africa, Zambia, Uganda and Ethiopia are consistent surplus producers; Malawi, Mozambique and Tanzania are either exporters or importers depending on weather conditions. Yet other countries, such as Kenya and Zimbabwe, have steadily increased imports in recent years and depended on imports for as much as 27% of domestic consumption in 2015-17.

Most trade occurs within the region. Trade policies tend to prioritise a stable supply for domestic markets, e.g. by imposing export controls during perceived production shortages. Such restrictions often limit access to local and regional supplies, amplify price swings, and add to import costs as countries have to source supplies internationally.

In the coming decade, white maize will continue to play a pivotal role for the region's food security (Table 1.1). The *Outlook* foresees further increases in food demand as growing per capita consumption of maize combines with strong population growth. This is expected to result in 18.4 Mt of additional maize food use over the coming decade, about half of the global growth in food consumption for maize.

Productivity growth among regional suppliers is key to ensuring progress towards the Zero Hunger target. Additionally, open and reliable trade relationships are crucial to sustain food security. Sub-Saharan Africa will be increasingly dependent on imports from other regions, as not all of the rising demand can be satisfied through local production.

Table 1.1. Per capita calorie availability for maize versus other food products

	2015-17		2027	
	Calories per capita	Share of total	Calories per capita	Share of total
Maize	491	19%	515	19%
Other cereals	784	30%	827	31%
Other crops	530	20%	536	20%
Animal products	188	7%	194	7%
Sugar	130	5%	137	5%
Vegetable oil	217	8%	235	9%
Other	255	9%	268	10%
Total	2 596	100%	2 711	100%

Note: Data refers to the average value for Sub-Saharan Africa.
Source: OECD/FAO (2018), "OECD-FAO Agricultural Outlook", OECD Agriculture statistics (database), http://dx.doi.org/10.1787/agr-outl-data-en.
1. This box summarises a more extensive analysis of the white maize market in Sub-Saharan Africa, available at www.agri-outlook.org.

Meat and fish: Global convergence in consumption patterns remains limited

Compared with cereals, which are an important food source across the world, consumption of meat and fish differs significantly across regions according to dietary patterns and income levels (Figure 1.6). The availability of meat and fish is particularly low in Sub-Saharan Africa, where low incomes limit consumption, and in India, where dairy constitutes an important part of protein intake. Availability is high in advanced economies and in Latin America (not shown on the chart), but also in China, where fish and pig meat account for more than half of the total.

At a global level, total consumption of meat and fish is expected to increase by 15% over the outlook period, while per capita consumption of meat and fish rises by only 3%, with stark variations across regions (Figure 1.6). The strongest growth in total consumption is expected in Sub-Saharan Africa (+28%), although this reflects exclusively the impact of population growth; per capita consumption is expected to decline by 3%. By contrast, per capita consumption growth is higher in India (+12%, albeit from a low base) and China (+13%).

For meat, per capita consumption will grow most strongly in absolute terms in the developed world (+2.9 kg/capita over the outlook period), facilitated by lower prices. A growing gap thus exists with developing countries, which expand availability by 1.4 kg/capita. This smaller expansion is partly a reflection of income constraints, supply chain issues in some areas (e.g. lack of a cold chain infrastructure) and, in some regions, dietary preferences where protein is obtained more from non-meat sources. Within the developing world, least developed countries will add only 0.3 kg/capita, due to slow growth in disposable income. Asian countries in this group are projected to show some growth while Sub-Saharan Africa is expected to experience declining per capita consumption of both meat and fish.

Figure 1.6. Meat and fish: Per capita availability for food consumption

(a) Per capita food consumption of meat and fish, by region and commodity, 2027

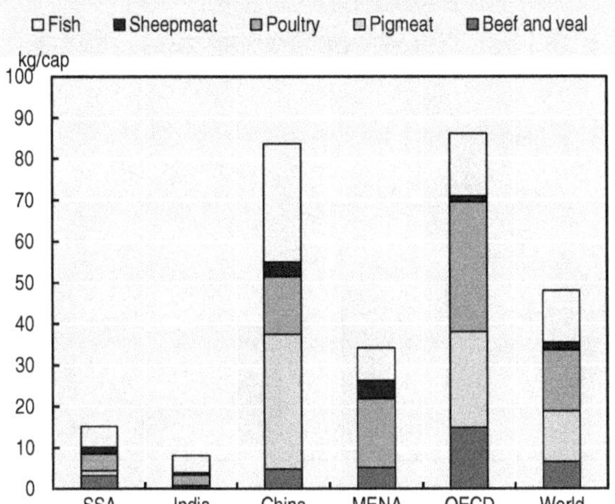

(b) Growth rates of per capita and total meat and fish consumption over the outlook period

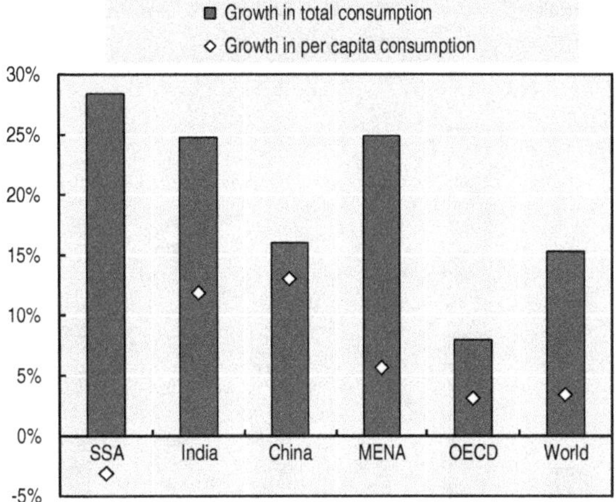

Note: SSA is Sub-Saharan Africa; MENA is Middle East and North Africa. Consumption is defined here in terms of food availability, and hence does not account for waste. Per capita consumption data refers to edible weight, estimated using conversion factors of 0.7 for beef and veal; 0.78 for pigmeat; 0.88 for poultry and sheep; and 0.6 for fish.

Source: OECD/FAO (2018), "OECD-FAO Agricultural Outlook", OECD Agriculture statistics (database), http://dx.doi.org/10.1787/agr-outl-data-en.

StatLink http://dx.doi.org/10.1787/888933741998

The past decade saw strong growth in the global per capita consumption of poultry (+16%), while the per capita consumption of beef and veal decreased by almost 5% between 2008 and 2017. For the coming decade, per capita consumption of poultry (typically the least expensive meat) is expected to increase by 5.5%, while beef and veal is projected to recover, with growth of 3.5% over the next decade, notably in China. Per capita pigmeat consumption will be flat at the global level, but growth is expected to be strong in regions and countries where pork is popular, such as Latin America and the Philippines, Thailand and Viet Nam. The role of China in global pork consumption growth is anticipated to diminish due to an already-high level of per capita consumption. Whereas China accounted for 65% of the increase in the previous decade, it will only contribute 45% of the expansion in the next ten years. Sheepmeat will remain a niche market in most countries, despite per capita consumption growth of 8% over the next ten years, concentrated mostly in China and other Asian countries as diets in the region diversify.

Dairy: Consumption of fresh dairy products expands in emerging economies

Dairy products can be consumed as fresh dairy products, butter, cheese, or as milk powders (e.g. for use in food processing). Fresh dairy products dominate consumption in developing regions and at a global level, while processed products such as butter and cheese dominate dairy consumption in the developed world (panel (a) (Figure 1.7).

Figure 1.7. Global consumption of dairy (in milk solids)

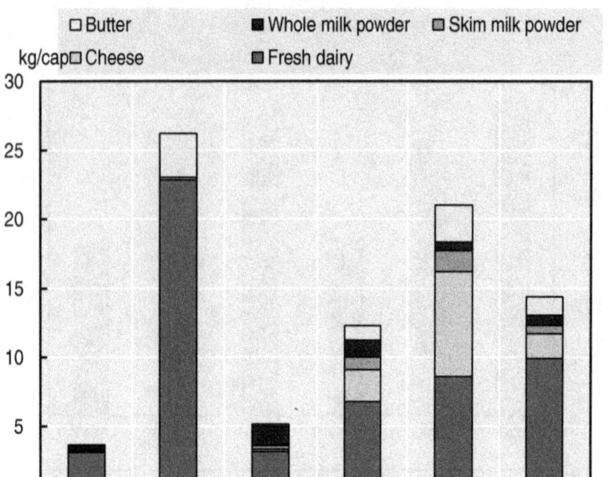

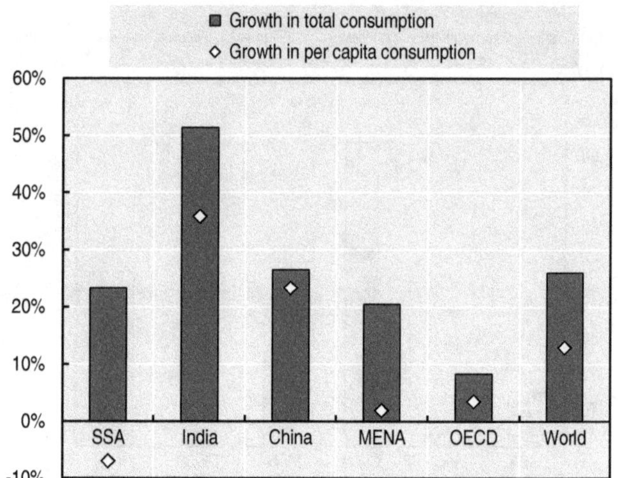

Note: Food consumption of dairy products in milk solid (fat and non-fat solid) equivalents. SSA is Sub-Saharan Africa; MENA is Middle East and North Africa. The Agricultural Outlook measures consumption in terms of food availability and hence does not account for waste.

Source: OECD/FAO (2018), "OECD-FAO Agricultural Outlook", OECD Agriculture statistics (database), http://dx.doi.org/10.1787/agr-outl-data-en.

StatLink http://dx.doi.org/10.1787/888933742017

The dominance of fresh dairy products will increase in the coming decade, with 2.2% p.a. growth in consumption, the highest growth rate among the commodities covered in the *Agricultural Outlook*. This increase can be attributed largely to India, where dairy is an integral component of the diet. In Ukraine and Kazakhstan, per capita consumption is also expected to grow strongly from already-high levels.

While developing countries are increasingly consuming fresh dairy products, adding 8.4 kg/capita by 2027, fresh dairy consumption in developed countries will fall by 1.7 kg/capita as consumers continue shifting towards processed dairy products, such as milk powders, cheese and butter.

A growing preference for butter in higher-income countries has been attributed in part to changing perceptions of the health implications of consuming dairy fat. Despite strong price movements in the past year, global demand for butter is expected to grow at nearly 2.2% per year. This growth will be supported by high and expanding consumption in India.

Sugar and vegetable oil: Consumption rising despite mounting health concerns

In addition to fresh dairy products, relatively high growth rates are also expected for sugar and vegetable oil, as urbanisation in developing countries leads to greater demand for convenience foods, typically characterised by a higher sugar and oil content.

Most of the additional demand for sugar will originate from the developing world (94%), in particular Asia (60%) and Africa (25%), two sugar-importing regions. Per capita consumption is projected to grow by 2.4 kg/capita in India, 2.5 kg/capita in China and 2.9

kg/capita in the Middle East and North Africa, compared with flat consumption in developed countries (Figure 1.8). In Sub-Saharan Africa, per capita consumption is projected to increase by 7% or 0.8 kg/capita over the next decade. Combined with strong population growth, total consumption in the region is expected to grow by 42%. While the increase in per capita consumption in Sub-Saharan Africa is relatively small, it contrasts with the projected decline in per capita consumption of meat, fish and dairy.

Figure 1.8. Food consumption of sugar

(a) Per capita food consumption of sugar

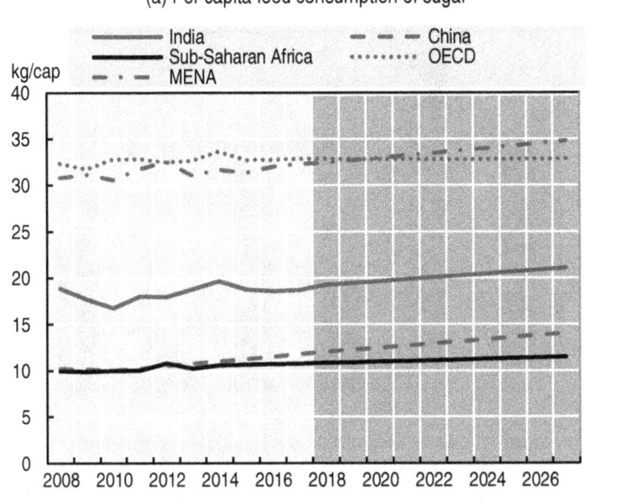

(b) Growth rates of per capita and total consumption over the outlook period

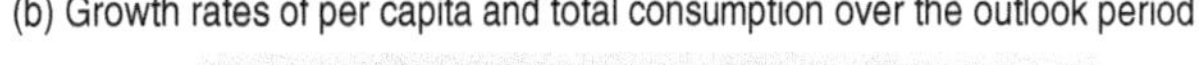

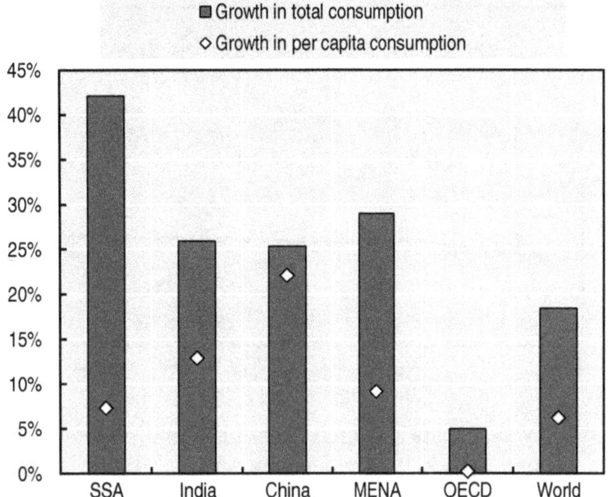

Note: Charts show food consumption of sugar from sugarcane and sugar beet (i.e. excluding other sweeteners such as high-fructose corn syrup). SSA is Sub-Saharan Africa; MENA is Middle East and North Africa. The Agricultural Outlook measures consumption here in terms of food availability and hence does not account for waste.

Source: OECD/FAO (2018), "OECD-FAO Agricultural Outlook", OECD Agriculture statistics (database), http://dx.doi.org/10.1787/agr-outl-data-en.

StatLink http://dx.doi.org/10.1787/888933742036

As for other commodities, patterns of sugar consumption are influenced by local factors as well as by incomes and preferences. For instance, per capita consumption is high in Brazil (the world's largest sugar producer) and other Latin American countries, and projected to continue increasing. Per capita consumption levels are also high in OECD countries, but projected to remain flat. This stagnation may partly be due to the identification of high levels of sugar consumption as a contributory factor to rising rates of obesity and non-communicable diseases. By contrast, even though per capita consumption levels in the Middle East and North Africa are similar to those in OECD countries, those factors are not expected to limit sugar consumption over the next ten years, which will continue to rise.

Compared with other commodities, expected growth in food demand is strong for vegetable oil, at 2.0% per year, although this represents a considerable slowdown compared with last decade's 3.9% annual growth rate.

For the world as a whole, per capita food use of vegetable oil is projected to increase from 21 to 23 kg per capita (Figure 1.9). In several developing countries, per capita consumption is approaching levels seen in the developed world. This is especially true for China, but also for India and the Middle East and North Africa. By contrast, per capita consumption in Sub-Saharan Africa will remain at levels much below those in the rest of the world, although it is projected to increase by 6% over the outlook period, or 0.6 kg/capita.

Figure 1.9. Food consumption of vegetable oil

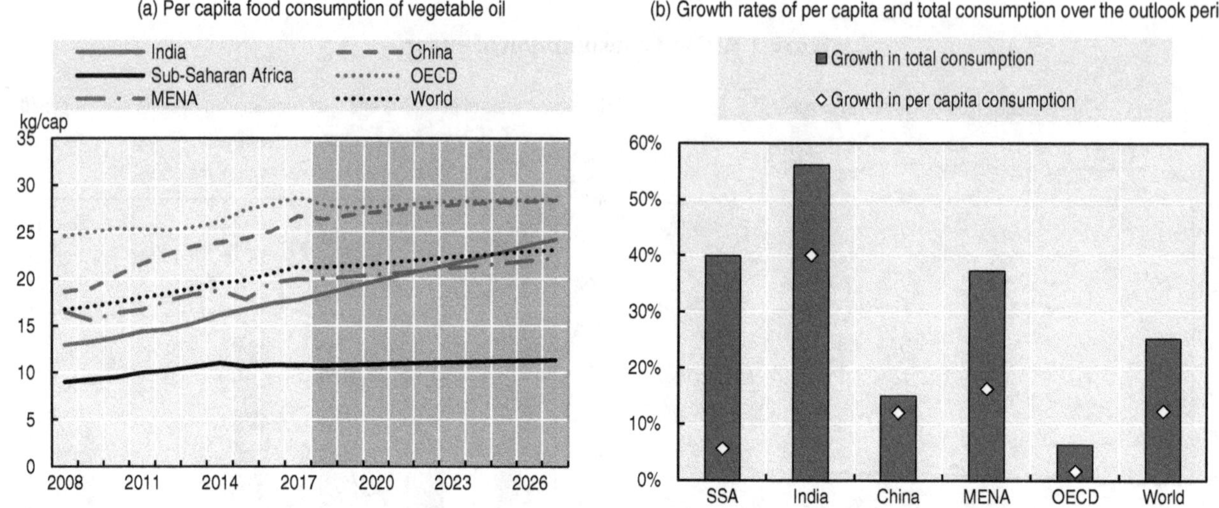

Note: Charts show food consumption of vegetable oil (i.e. excluding use as feedstock for biodiesel and other uses). SSA is Sub-Saharan Africa; MENA is Middle East and North Africa. The *Agricultural Outlook* measures consumption here in terms of food availability and hence does not account for waste.
Source: OECD/FAO (2018), "OECD-FAO Agricultural Outlook", OECD Agriculture statistics (database), http://dx.doi.org/10.1787/agr-outl-data-en.

StatLink http://dx.doi.org/10.1787/888933742055

As the preceding discussion shows, the strong demand growth in the developing world does not always correspond to increasing per capita availability of food. In Sub-Saharan Africa, high growth rates for fish and meat are the result of strong population growth, while per capita availability is expected to fall; while in the Middle East and North Africa, per capita availability of meat and fish is not expected to increase much. By contrast, in these regions per capita availability of sugar and vegetable oil are expected to increase. More generally, Least Developed Countries (LDCs) are expected to increase their calorie availability at a slower rate in the coming decade, and this increase is due mostly to increased sugar and oil consumption while per capita intake of animal proteins is expected to remain low. As a result, malnutrition will remain an important problem in LDCs, as detailed in Box 1.2.

Box 1.2. Prospects for food consumption and nutrition in Least Developed Countries

The United Nations recognise Least Developed Countries (LDCs) as particularly disadvantaged and deserving of special international support. Currently, countries with an annual per capita income below USD 1 025, a low level of human capital, and a structural vulnerability to economic and environmental shocks are classified as LDC. Of these, 33 are located in Africa, 13 in Asia and the Pacific, and one in Latin America. They are home to 12% of the global population, but account for less than 2% of global GDP and only about 1% of global merchandise trade.

Economic conditions in several LDCs have improved over the last decade, as average per capita income growth in LDCs exceeded 3% per year. Subsequently, the Prevalence of

Undernourishment (PoU) in LDCs as a group fell from 32.8% in 2000-2002 to 23.8% in 2010-2012. However, estimates for 2014-2016 suggest a rebound to 24.4%, equivalent to 232 million undernourished people.

Conflict and weather-related production shocks have been identified as the main factors driving the recent rise in undernourishment, particularly in the Middle East and North Africa. Wars and civil strife have been disrupting domestic economic activities and foreign exchange earnings as well as damaging local food production. Food import dependency, particularly for cereals, remains high in several of the most food insecure LDCs. Countries that are simultaneously affected by conflict and climate-related shocks have seen a particularly large toll on their food security. In 2016, these factors severely compromised the food security of 45 million people in eight LDCs (Afghanistan, Burundi, Central African Republic, Democratic Republic of the Congo, Somalia, South Sudan, Sudan, and Yemen).

Figure 1.10. Sources of calories and proteins in Least Developed Countries

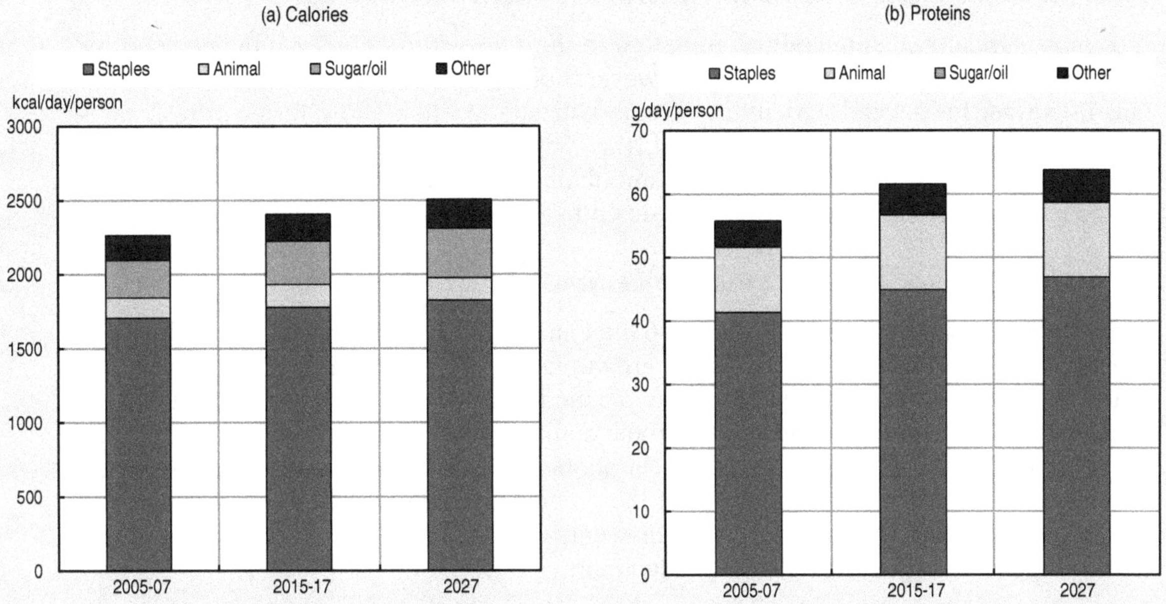

(a) Calories

■ Staples □ Animal ■ Sugar/oil ■ Other

kcal/day/person

3000, 2500, 2000, 1500, 1000, 500, 0

2005-07, 2015-17, 2027

(b) Proteins

■ Staples □ Animal ■ Sugar/oil ■ Other

g/day/person

70, 60, 50, 40, 30, 20, 10, 0

2005-07, 2015-17, 2027

Source: OECD/FAO (2018), "OECD-FAO Agricultural Outlook", OECD Agriculture statistics (database), http://dx.doi.org/10.1787/agr-outl-data-en.

StatLink http://dx.doi.org/10.1787/888933742074

The macroeconomic outlook for LDCs projects a 3% annual growth of per capita income over the next decade. This rate is expected to support a further increase in the calorie availability of LDCs, but at a slower rate. In the last decade, daily calorie availability grew 115 kcal to 2 415 kcal/day. In the coming decade, daily calorie availability in LDCs is projected to rise by 85 kcal, reaching 2 505 kcal/day by 2027. This is 30% lower than the projected level in developed countries, which should reach 3 482 kcal/day by 2027.

Not only has progress been limited in expanding calorie availability, it was also unevenly distributed across countries and regions – and will continue to be so. In Asian LDCs, calorie availability is estimated to reach almost 2 700 kcal/day, while African LDCs are expected to reach only 2 450 kcal/day by 2027 despite a higher growth rate. Food availability in LDCs in the Middle East and North Africa fell in recent years, but it is expected to recover from an current average of 2 270 kcal/day to 2 420 kcal/day by 2027.

Staples (cereals, pulses, roots and tubers) are expected to remain the main source of calories in LDCs, even if their share is expected to gradually decline to 73% in 2027, down from 75% in 2005-2007. The additional dietary energy is expected to come from more sugar and fats, which are predicted to increase their share from 12% in 2015-2017 to 13% in 2027.

Even less progress is expected in improving the protein intake. Average protein availability will remain about 64 grams per day in 2027, mostly from cereals, with the availability of high-quality animal proteins reaching only about 12 grams per day. Consumers in LDCs will continue to have access to only a limited variety of foods and therefore their diets will still lack macronutrient diversity and essential micronutrients, adding to the burden of persistent calorie deficits.

The slow growth in dietary energy and continued poor nutrition prospects also suggest that many LDCs will not be able to meet the UN's Sustainable Development Goal of eliminating all forms of malnutrition by 2030. Achieving this goal would require substantial progress in reducing conflicts while helping smallholders to improve local production and bring about resilience to climate change and weather-related shocks.

Non-food uses affect demand for several agricultural commodities

For most agricultural commodities reviewed in the *Agricultural Outlook*, the demand for food uses dominates overall demand. However, non-food uses, particularly feed and fuel, are important for several agricultural commodities, and often show faster growth rates than food demand. In the case of feed, this will remain true in the coming decade. Biofuels by contrast were a major factor stimulating demand for agricultural commodities in the past decade, but growth is slowing down in the coming decade.

Feed: Rising share of global crop output directed towards feed use

The global demand for feed reached 1.6 bln t in 2015-17, and is expected to increase further to 1.9 bln t by 2027, at an annual growth rate of around 1.7%. Demand for feed is thus expected to grow faster than the demand for several commodities shown in Figure 1.11 and markedly faster than food demand for cereals, for which 1.1% p.a. growth is expected. This growth results in about 260 Mt of additional feed demand by 2027; slightly less than the expansion of the previous decade in which demand grew by more than 300 Mt. Demand for feed also outpaces the growth in demand for meat, indicating an intensification of meat production.

The main set of agricultural commodities used for feed includes maize, protein meal, other coarse grains (especially barley and sorghum), wheat, and by-products of cereal processing such as cereal bran. As shown in Figure 1.11, maize and protein meal will remain the most important commodities used as feed, accounting for 60% of all feed by 2027 (up from 58% in the base period). Feed demand for maize is expected to grow by 21% over the outlook period, and demand for protein meal is expected to expand by 23%, considerably faster than the other commodities used as feed.

Projections for protein meal, which is derived from crushing oilseeds, will be influenced by developments in feed systems and in agricultural policies. For instance, Least Developed Countries' total demand is expected to grow around 45% between 2015-17 and 2027, reflecting the intensification of livestock production as these countries move towards compound feed-based livestock production. Yet, global demand growth for protein meal is expected to fall below the average annual rate of the past decade (1.7% compared to 4.2%). That high growth rate was in large part due to China, where the intensification of meat production coincided with a high support price for grains. This discouraged the use of maize as feed. The reduction of maize support prices in China since 2016 means that maize will play a more important role in the Chinese feed mix in the next decade.

Overall growth patterns in the demand for feed will vary across geographic regions. Around 30% of the additional demand for feed will originate in China, where feed demand is expected to grow 25% over the outlook period. Strong growth in feed demand is also expected in the Middle East and North Africa (+29%, with the region expected to account for around 10% of additional global demand), as well as Brazil (+25%) and India (+31%). Growth rates in the European Union and the United States are considerably lower at 0.4% and 11% over the outlook period respectively. For the European Union, this rate reflects the expected decline in domestic meat consumption over the outlook period.

Figure 1.11. Demand for feed

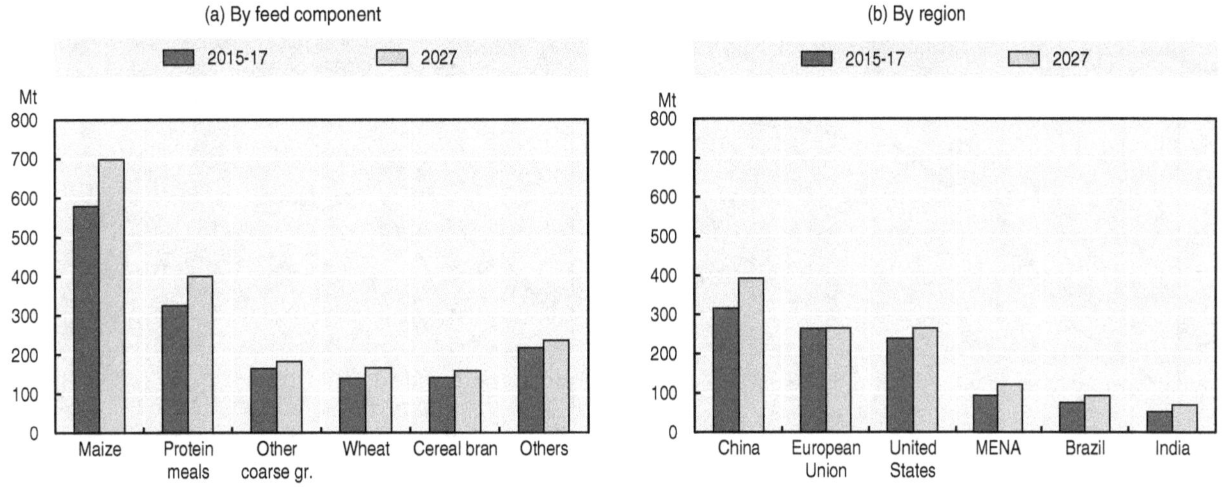

Note: MENA stands for the Middle East and North Africa.
Source: OECD/FAO (2018), "OECD-FAO Agricultural Outlook", OECD Agriculture statistics (database), http://dx.doi.org/10.1787/agr-outl-data-en.

StatLink http://dx.doi.org/10.1787/888933742093

Fuel: Growth in Brazil and emerging producers

Agricultural commodities are not only used as food and feed, but also as fuel in the form of biofuels. These include ethanol, based mostly on maize and sugarcane, and biodiesel, produced mostly from vegetable oil. The evolution of biofuels is highly sensitive to potential changes in policy, as well as to overall demand for transport fuel, which in turn depends on the crude oil price. In many countries, mandatory blending rules impose a minimum share of ethanol and biodiesel to be used in transport fuel. The link between oil prices and biofuel prices is therefore complex, as explained in more detail in Box 1.3 The baseline projections in the *Outlook* are based on the policies currently in place in the key regions. Projections are clearly sensitive to changes in that policy environment.

In the second half of the 2000s, various policies started to stimulate biofuel production, leading to a strong increase in world ethanol and biodiesel output. As a result, a growing share of global sugarcane and maize production was used for ethanol production, while a growing share of vegetable oil was used for biodiesel production (Figure 1.12). This policy-induced expansion of biofuels was a major driver of increased demand for maize, sugarcane and vegetable oil over the past decade.

Over the next ten years, the demand for these commodities as inputs to biofuel production is expected to stabilise, as mandatory blending requirements are not expected to rise at the same pace as over the past ten years. As such, the production of biofuels is expected to grow more slowly over the coming decade. In the past ten years, global production of ethanol grew by 64 billion litres (bln L), equivalent to 3.9% p.a. growth; over the next ten years, only an addition of 12 bln L (0.7% p.a.) is projected. For biodiesel, the past decade saw an increase of 29 bln L (9.5% p.a.), whereas only 5 bln L (0.4% p.a.) is expected to be added over the outlook period.

Figure 1.12. Biofuels and the demand for feedstock, 2000-2027

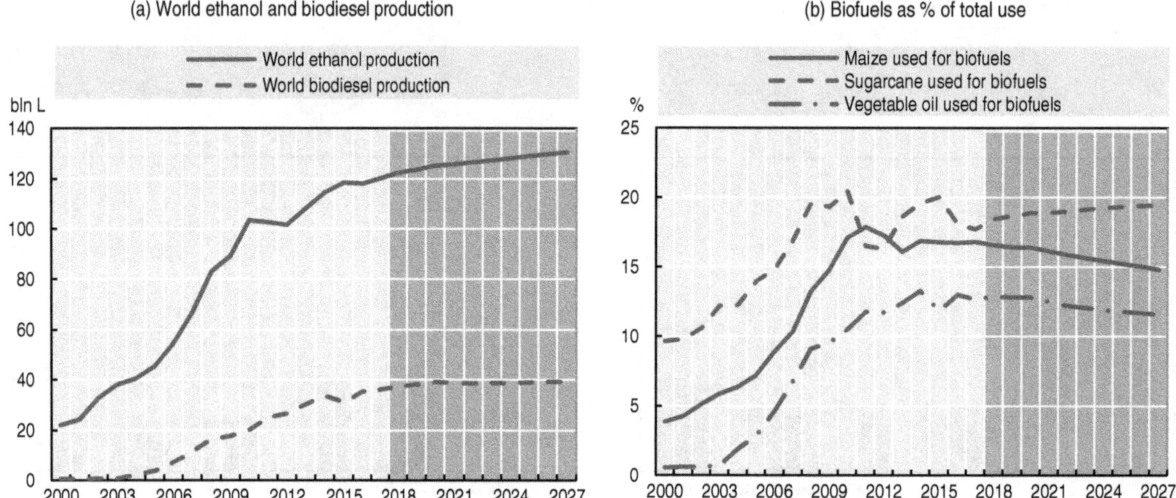

Source: OECD/FAO (2018), "OECD-FAO Agricultural Outlook", OECD Agriculture statistics (database), http://dx.doi.org/10.1787/agr-outl-data-en.

StatLink http://dx.doi.org/10.1787/888933742112

The composition of the demand for biofuels is also changing, with a shift towards developing countries which are increasingly putting in place policies favouring the domestic biofuel market. For ethanol, the main markets are the United States, Brazil, China, and the European Union. Declining demand for transport fuel is expected to decrease the demand for ethanol in the United States and the European Union, while strong growth is expected in Brazil, China, and Thailand, spurred by favourable policies. The demand in China could increase further with the implementation of the country's proposed new ethanol mandate (discussed in the Biofuels chapter). Overall, 84% of the additional demand for ethanol in the coming decade will come from developing countries.

For biodiesel, the main markets are the European Union, the United States, Brazil, Argentina and Indonesia. As with ethanol, demand is expected to decline in the European Union and the United States, which will drive down demand for vegetable oil as feedstock. Instead, an expansion is expected in Brazil, Argentina, Indonesia, and other developing countries, again mainly through favourable policy measures.

Food, feed and fuel: Competing sources of cereal demand

In addition to being an important and relatively low-cost source of calories, cereals are widely used for feed and fuel, in large part because of the ease with which cereals can be processed into other forms. This versatility also implies that food use of cereals may come into competition with non-food uses, especially when non-food uses expand rapidly.

As shown in Figure 1.13, between 2005-7 and 2017 the global demand for cereals increased by around 520 Mt to around 2.6 bln t. Over the coming decade, demand will grow by around 360 Mt, but the composition of this demand growth is changing. While fuel was a major component of demand growth in the past decade (contributing more than 120 Mt to demand), this is no longer expected to be the case over the outlook period. Instead, food and feed uses are driving growth, together accounting for almost all additional demand over the coming decade.

Panel (b) shows cereal demand by crop. In the past decade, maize accounted for almost 330 Mt of the 520 Mt of additional cereals demand, or more than 60%. Over the outlook period, demand for maize will grow by 164 Mt, accounting for 46% of demand growth only. This slowdown in growth is consistent with the evolution of biofuel markets over the coming decade. For both rice and wheat, demand growth is expected to be more robust, with 97 Mt of additional wheat demand and 66 Mt of additional rice demand, most of it for food uses. Following flat demand over the last decade, renewed interest is expected in other coarse grains, which are projected to grow by more than 32 Mt over the coming decade. The projected trends in cereals are thus a reflection of the demand trends in food, feed and fuel.

Figure 1.13. Global demand for cereals, 2008-2027

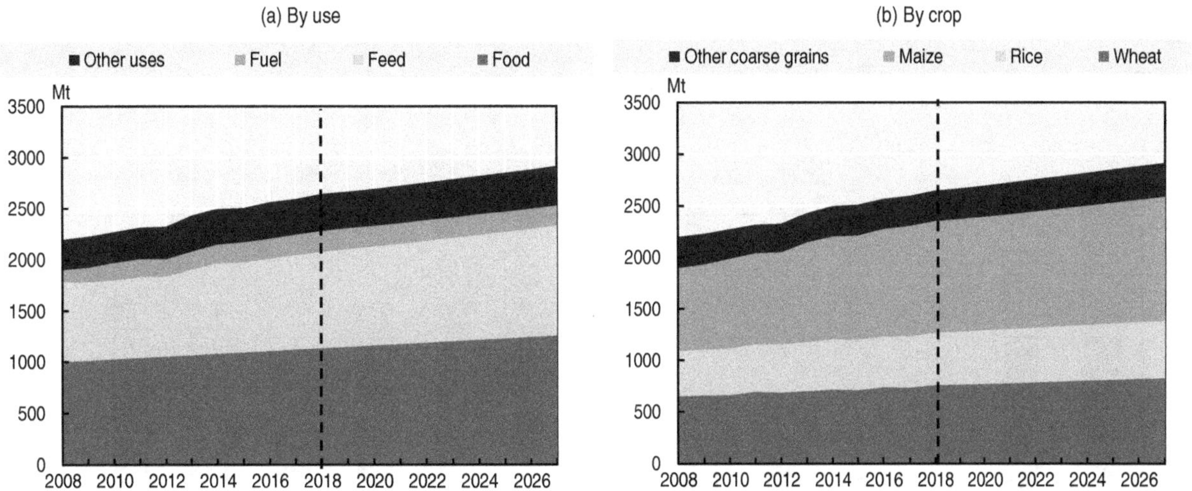

Note: The *Agricultural Outlook* measures demand in terms of availability, and hence does not account for waste.

Source: OECD/FAO (2018), "OECD-FAO Agricultural Outlook", OECD Agriculture statistics (database), http://dx.doi.org/10.1787/agr-outl-data-en.

StatLink http://dx.doi.org/ 10.1787/888933742131

Production

While the last decade was characterized by robust demand and high agricultural prices, leading to strong production growth across commodities, the coming decade will see global agricultural production grow more slowly. Under the current set of assumptions, agricultural and fish production is expected to grow by 1.5% p.a. over the coming decade, or a total growth of 16% over the outlook period. Most of this growth will be due to increasing productivity, with no major increase in agricultural land use at the global level, although this varies by commodity and by region. Trends across main producing regions are discussed in more detail below.

Agricultural output will grow with little change in global land use

Land is an important input into agricultural production, both for arable crops and for grazing. Production growth in agriculture can come from taking more land into production, or from increasing the output per unit of land. Since land use is largely defined according to agro-ecological characteristics, the availability of agricultural land and the relative share of cropland versus pasture differ strongly across regions (Figure 1.14). Since 1960, global agricultural land is estimated to have increased by around 10%, with most of this increase occurring before 1990 and relative stability since. At a global level, this relative stability is expected to continue over the coming decade.

Pasture land, used for grazing ruminants such as cattle, sheep, or goats, is mainly concentrated in three regions: the Americas, which hold over one-fourth of the world's pasture land; Sub-Saharan Africa, which accounts for 21% of global pasture; and South and East Asia, home to 17% of global pasture. While the Americas and South and East Asia also lead in global ruminant meat output, jointly producing more than 60% of global supplies in 2015-17, Sub-Saharan Africa only contributes about 8%. This low share is indicative of the small-scale and in large part traditional nature of the sector. Western Europe, by contrast, reports the lowest share of global pasture at 2%, yet accounts for 11% of global ruminant meat in 2015-17, an indication of the industrial nature of meat production in the advanced economies of that region.

Figure 1.14. Land use in global agriculture, 2015-17 and 2027

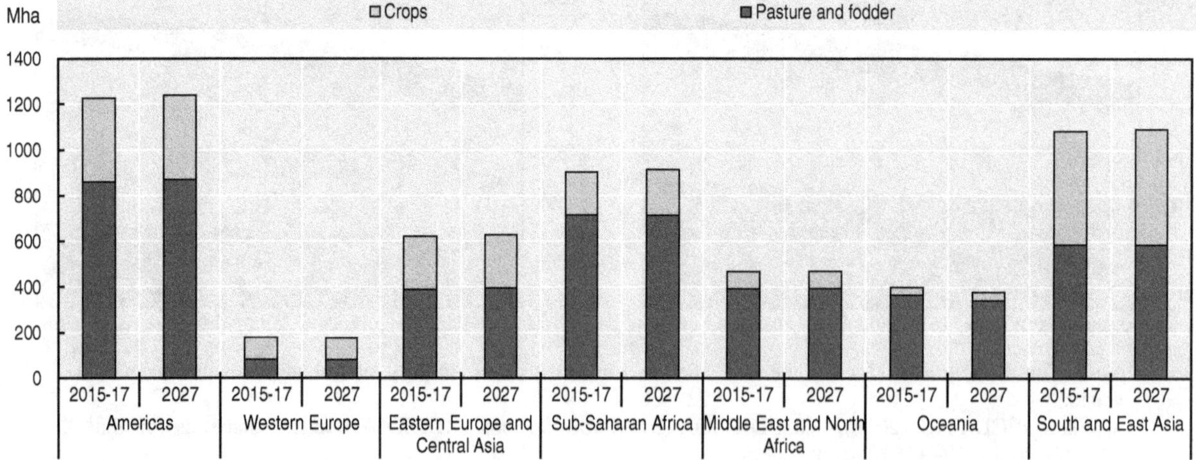

Note: Western Europe includes the European Union, Norway, and Switzerland; Eastern Europe and Central Asia includes the Russian Federation, Ukraine, Kazakhstan, Turkey, Israel, some smaller non-EU countries in Eastern Europe, and some smaller Central Asian countries); Middle East and North Africa is as defined in Chapter 2; Oceania includes Australia, New Zealand and some smaller countries in the region; South and East Asia includes all other countries.

Source: OECD/FAO (2018), "OECD-FAO Agricultural Outlook", OECD Agriculture statistics (database), http://dx.doi.org/10.1787/agr-outl-data-en.

StatLink http://dx.doi.org/10.1787/888933742150

Changes in ruminant meat production will not be accompanied by corresponding shifts in global pasture area over the outlook period. While global production is projected to increase 16% for beef and veal and 21% for sheepmeat, largely due to increases in output in the Americas, South and East Asia and Sub-Saharan Africa, the land allocation to pasture remains largely unchanged. Moreover, the non-ruminant meat sector, which does not require pasture, is also set to expand over the coming decade, with global poultry and pigmeat production growing 18% and 11% respectively.

About half of global crop land is dedicated to cereals and oilseeds. Given arable land constraints, the overall area of crop land is not expected to change substantially over the coming decade, and productivity growth will be essential for sustaining crop production growth. Nevertheless, changes in area allocation and yields will vary across crops and regions. For maize and other cereals, most of the production increase will come from higher yields, not from greater land use (except for maize in Latin America). For other crops, soybeans in particular, land use will play a greater role, as area expansion and greater cropping intensity is expected in Latin America (Brazil, Argentina) (Figure 1.15).

Yields are set to rise fastest for Sub-Saharan Africa, albeit from a low base, with high growth rates across practically all crops. This trend is indicative of the production potential of the region, but also of the relatively low yields experienced today for most major commodities. In comparison, Western Europe and the Americas will record more moderate yield growth rates, as productivity is already high for most crops. Figure 1.16 indicates that maize yields will reach 8.0 t/ha in Western Europe by 2027, and in the Americas 8.6 t/ha, compared to only 2.5 t/ha in Sub-Saharan Africa.

Figure 1.15. Pasture and ruminant meat production by region

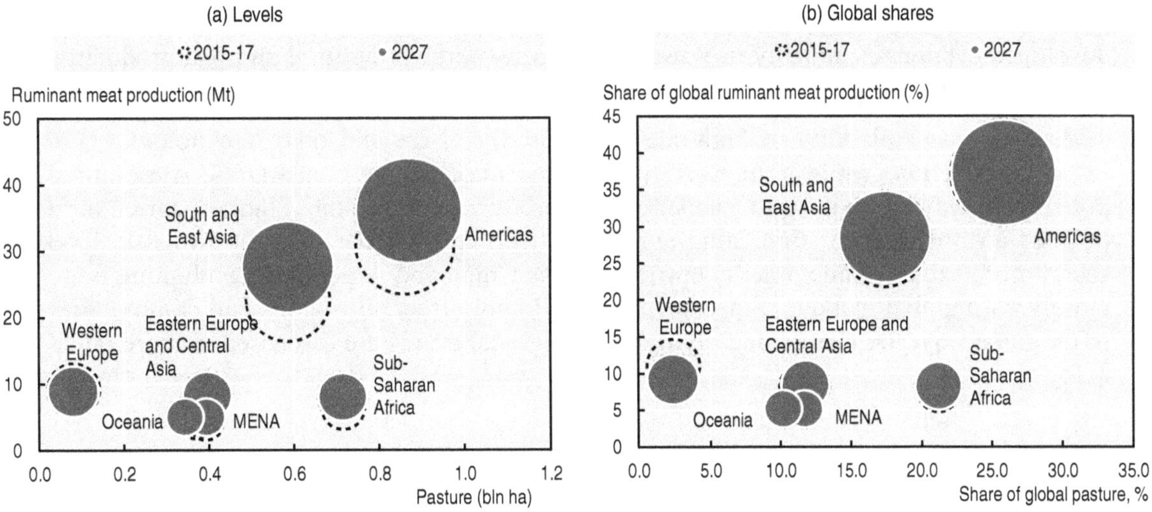

Note: MENA stands for Middle East and North Africa. The size of e ach bubble is proportional to the region's ruminant meat production level.
Source: OECD/FAO (2018), "OECD-FAO Agricultural Outlook", OECD Agriculture statistics (database), http://dx.doi.org/10.1787/agr-outl-data-en.

StatLink http://dx.doi.org/10.1787/888933742169

Figure 1.16. Crop land and yield trends for maize and soybeans

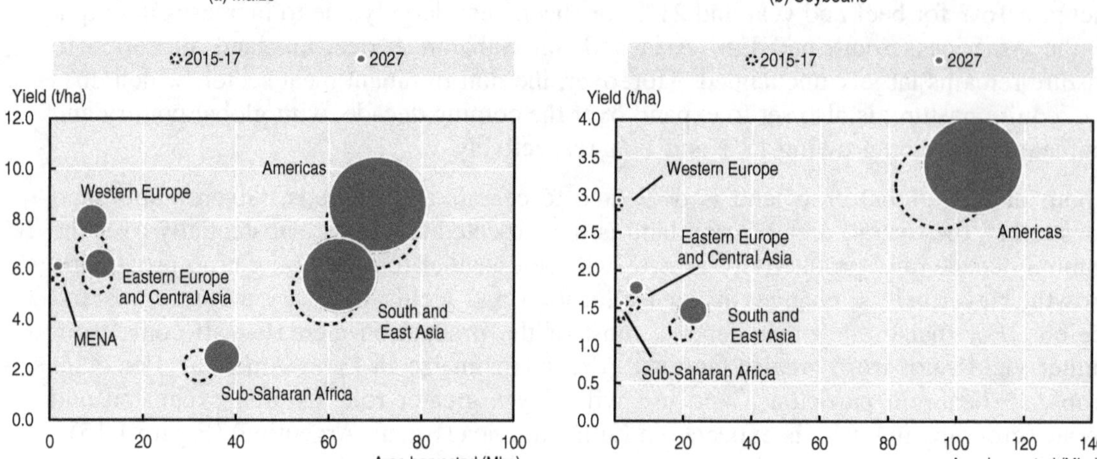

Note: MENA stands for Middle East and North Africa. The size of each bubble is proportional to the region's crop production.
Source: OECD/FAO (2018), "OECD-FAO Agricultural Outlook", OECD Agriculture statistics (database), http://dx.doi.org/10.1787/agr-outl-data-en.

StatLink http://dx.doi.org/10.1787/888933742188

Developing regions expand and intensify agricultural production

Over the coming decade, the expansion of agricultural production will be disproportionately concentrated in the developing world (Figure 1.17). The fastest growth is expected in Sub-Saharan Africa and South and East Asia, with the latter also expected to show the greatest growth in absolute terms. Overall, output will expand less in developed economies, notably in Western Europe, where agricultural and fish production is only projected to grow by around 3% over the outlook period.

The improved availability of high-quality seeds, fertilisers and other technologies will favour production, while sustainability concerns may impose constraints. Agricultural policies worldwide will also shape global production decisions. India's agricultural policies are focused on stimulating agricultural growth in order to meet domestic food security objectives, while other countries such as China and Argentina are aligning more closely with global markets. Since such trends do not affect all regions and commodities in the same way, the factors underlying these regional trends are discussed in more detail below.

Figure 1.17. Regional trends in production

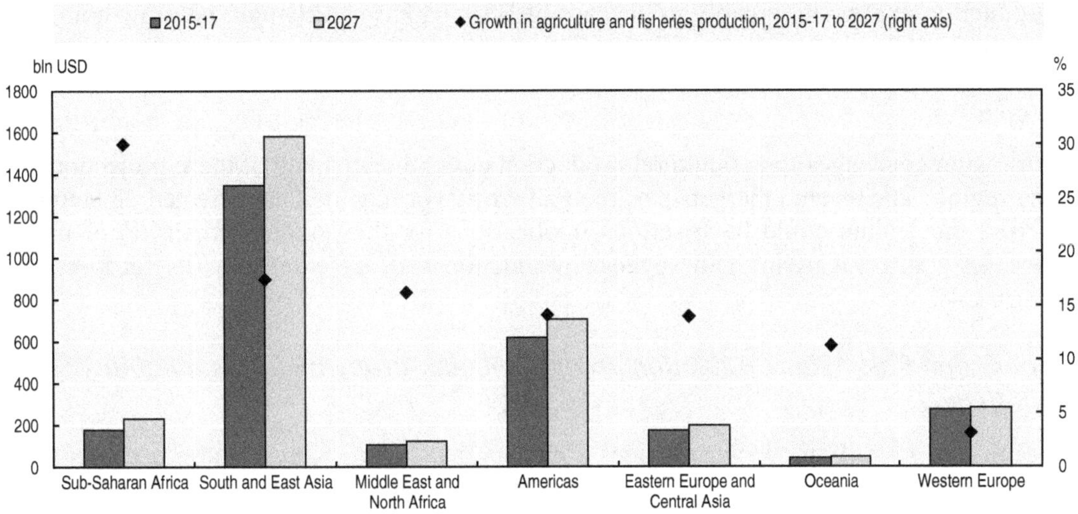

■2015-17 □2027 ◆Growth in agriculture and fisheries production, 2015-17 to 2027 (right axis)

bln USD | %

Sub-Saharan Africa, South and East Asia, Middle East and North Africa, Americas, Eastern Europe and Central Asia, Oceania, Western Europe

Note: Figure shows the estimated net value of agricultural and fish production, in billions of USD, measured at constant 2004-6 prices. Regions as defined in Figure 1.14.
Source: OECD/FAO (2018), "OECD-FAO Agricultural Outlook", OECD Agriculture statistics (database), http://dx.doi.org/10.1787/agr-outl-data-en.

StatLink http://dx.doi.org/10.1787/888933742207

Sub-Saharan Africa: Productivity gains in basic foods

Despite accounting for over 13% of the world population and close to 20% of global agricultural land, Sub-Saharan Africa's share of global agricultural output is relatively low. Agricultural production is constrained by challenging agro-ecological conditions, limited access to and utilisation of technology, and the fact that economic growth in many cases remains only marginally ahead of population increases. Its most important commodity, among those analysed by the *Outlook*, is 'other coarse grains' (including millet, sorghum and teff) for which Sub-Saharan Africa today accounts for 14% of global production.

However, robust growth in agricultural production is expected for the coming decade. Crop production will expand by 30%, while meat, dairy and fish will respectively grow by 25%, 25% and 12%. This output growth will be accompanied by area expansions for maize, soybeans, and sugarcane and improved productivity across the board. Fertiliser, pesticides, improved seeds, and other technologies such as mechanisation and irrigation have the potential to introduce substantial productivity gains, as adoption is typically low among the small-scale production units that characterise the region.

Even with the projected strong growth, the region's food security will continue to depend on global markets as domestic production capacity will remain insufficient to meet the region's growing consumption needs. At the same time, there are some commodities for which African countries have become regional providers. Maize is one example, where Zambia consistently produces an exportable surplus. Vegetable oil is another example, as West African countries seek to promote their palm oil sectors, and production is expanding rapidly, notably in Nigeria. Yield improvements are expected to contribute to a 22% growth in output of palm oil in Sub-Saharan Africa over the outlook period. Similarly, yield improvements in teff production in Ethiopia will allow the country to account for nearly one-fifth of the world's production growth in other coarse grains.

Robust production growth is also expected for cotton (+33% over the outlook period), sugarcane (+18%) and sugar (+34%). Yield improvements will account for the cotton production expansion, notably in the case of Burkina Faso. Although the growth in sugar and sugarcane production marks the region among the fastest growing for these two commodities, by 2027 Sub-Saharan Africa will still account for less than 5% of global output.

Emerging challenges to agricultural production could threaten any of these projections for the region. The recent emergence of the Fall Armyworm, which has affected 28 countries across the region, could have serious implications for the region's expansion of maize, rice, sorghum, sugarcane and soybean production and, by extension, its food security (Box 1.4).

South and East Asia: Production growth remains strong amid sustainability challenges

South and East Asia (which includes China, India, Japan, Korea, and the countries of Southeast Asia) is the world’s main producer for an array of agricultural products. Despite facing serious constraints in terms of land, water, and workforce shortages, the region produces almost 40% of the world’s output of cereals (including nearly 90% of global rice output); close to 40% of global meat production; more than half of vegetable oil supplies; and nearly 70% of combined global capture and aquaculture fish production.

The coming decade is likely to add new challenges, in particular the need to reconcile high output levels with increasingly stringent standards for sustainable production. Nonetheless, the region is expected to expand agricultural and fish production by 17% over the outlook period.

Yield improvements will underlie a large part of the expansion of crop production, with yield growth over the outlook period of 10% for wheat, 12% for maize and rice, 15% for cotton and 20% for soybeans. While these figures are in line with global trends, oilseed yields are set to rise strongly in India, led by investment in production and information technologies, such as eNAM, an online trading platform for agricultural commodities. Oilseed production and crush are anticipated to expand as India seeks to meet growing domestic demand for vegetable oil.

Indonesia and Malaysia will continue to supply most of the world’s palm-oil sourced vegetable oil. Intensification of production on existing palm oil plantations is foreseen as area expansion possibilities are limited, especially in light of global pressure to improve the sustainability of palm oil production.

South and East Asia will remain a major global supplier of meat and dairy commodities, accounting respectively for 39% and 44% of global output by 2027. Dairy production is set to expand by 41% over the outlook period with butter rising 44% and milk expanding 40%. Meat production will grow by 18%. China, India and Thailand will lead output growth of poultry and sheep meat, while slower growth in pork for the region is explained by a slowdown in China’s production.

Fish production from capture and aquaculture will expand by 15% in South and East Asia, despite China’s plans to scale back fisheries production over the coming decade and introduce sustainable practices to the sector. If China’s 13th five-year-plan is fully implemented, capture fisheries from China will contract by about 29% by 2027, and aquaculture will expand by 20%, instead of 31% in the absence of the plan. With limited global capacity to fill China’s production gap under the five-year-plan, upward pressure will be placed on global fish prices (as discussed in more detail in the Fish and seafood chapter).

Biofuels production expansion in the region is also led by China, which is expected to become the world's third largest ethanol producer, with output of 11 bln L by 2027. About half of that output will be directed to biofuel production; the remainder is allocated to industrial uses. This projection does not take into account the possible impact of a recently proposed new nationwide E10 ethanol mandate. If implemented, this could raise Chinese ethanol output to 29 bln L by 2027, similar to the expected output for Brazil. (The possible impacts of this mandate are explored in more detail in the Biofuels chapter). Thailand is also projected to play a prominent role in regional and global ethanol markets producing 3.2 bln L in 2027. In terms of biodiesel, Indonesia will continue to be the region's main producer (4.3 bln L in 2027).

In India, policy makers focus on promoting agricultural growth to meet domestic food security objectives, and policies will likely seek to stimulate investment in the domestic agricultural sector both through protection against import competition through import tariffs and through producer support. While India's policies may impact domestic production more than global markets, China's policies, in particular regarding cereals, are likely to affect global markets through price movements, stock releases and import regulations. The reduction in maize support prices since 2016 will have implications in the coming decade for domestic and global production of maize, soybean, and other coarse grains.

Middle East and North Africa: Improved economic growth set to stimulate agricultural output

The agricultural sector in the Middle East and North Africa has historically been constrained by unfavourable agro-ecological conditions for crop production as well as political instability. However, over the coming decade, the region is projected to enter a period of improved economic growth, which should underpin 16% growth in agricultural and fish output over the coming decade. Greater agricultural production will depend on innovation to enhance productivity growth in the face of scarcity of water and arable land across the region.

Livestock activities serve as the main source of agricultural value added in the region, with regional production of meat and dairy largely taking place in Iran and Egypt. Poultry is the main type of meat produced by these countries, each of which will pursue both extensive growth and productivity improvements over the coming decade. Production of milk, maize and oilseeds will expand at faster rates than over the previous decade. Nonetheless, the region will remain a net importer of these and most major commodities, given its various production constraints.

More details on the region's production trends can be found in Chapter 2, which offers an in-depth discussion of the agricultural sector with disaggregated projections for most countries in the region.

Americas: Export-oriented agricultural sectors respond to global demand

Alongside South and East Asia, the Americas are major producers of most commodities analysed by the *Outlook*. The region accounts for nearly 90% of global soybean production, and also holds large shares of global production in cereals (28%), specifically maize (52%). The region is a large producer of commodities with a high value added, such as protein meal, sugar, and biodiesel, where it accounts for 41%, 39%, and 42% of global output, respectively. With area harvested expanding and crop intensities rising over the next ten years, crop production in the region is anticipated to grow by 14%.

Area expansion will lead sugar production from Brazil, the world's leading producer, to grow 1.9% p.a., contributing to a 1.8% p.a. growth for the region as a whole. This growth occurs despite replanting setbacks and the competition between sugar and sugarcane-based ethanol production, as Brazil is also a global leader in biofuel production. Ethanol production in Brazil is set to expand by 1.5% p.a. over the outlook period. However, its global share will decline from 90% to 88%, given a rapid expansion of production in Asia.

Global soybean production will remain dominated by the United States and Brazil. In Brazil, higher cropping intensity will sustain its position, as it obtains soybean as a second crop on land cultivated with maize. These expansions will provide an input to regional dairy production and the global supply of protein meals and vegetable oils. In this context, Colombia is expected to become a net exporter of vegetable oil over the outlook period, expanding its area for oil palm cultivation, while Paraguay will follow trends in Brazil, expanding the area dedicated to soybean production and increasing oilseed crush.

The development of protein meal production will be essential for feeding the region's growing livestock sector. The United States and Brazil will continue to produce most of the world's meat supplies, with herd expansions in both regions. Production is expected to grow by 17% for beef and pork, 16% for poultry and 9% for sheep. Animal products such as milk and eggs will grow at similarly robust levels. Fish production is expected to expand by 9% over the outlook period, with a major expansion in aquaculture (+35%), in particular in Brazil and Chile.

Eastern Europe and Central Asia: Growing prominence in global cereals market

Agricultural production in Eastern Europe and Central Asia (a region which includes the Russian Federation, Ukraine, Kazakhstan and Turkey as main agricultural producers) expanded rapidly over the previous decade, due to an overall economic recovery and considerable investments into the modernisation of agriculture. In the coming decade, agricultural and fish production will expand by 14%.

In terms of arable crops, the region will maintain its position as second-largest wheat producer, increasing its share of global production to almost 22% by 2027. Maize output will also expand by 17% over the outlook period, although the region's global share will remain relatively low, at less than 6% by 2027. The region's share in global production of sunflowers and rapeseed will increase from 22% in 2015-17 to 25% by 2027, underpinned by an expansion in area harvested, which will be offset by a reduction in the area for roots and tubers.

These shifts in crop production are largely attributable to the evolutions in the Russian Federation, with area expansion underpinning growth in production of soybeans, other oilseeds, cereals and sugar beet. For the rest of the region, yield improvements will contribute disproportionately to output growth.

Livestock production will grow both in terms of meat and dairy products, accompanied by an increase in pasture area of 2% over the outlook period. The meat sector will expand 16% for the region, despite much slower output growth for the Russian Federation. Dairy production in the Russian Federation will be flat over the next ten years (following a contraction of 0.7% p.a. in the last decade). For the region as a whole, milk production will grow 1.1% p.a., and dairy processing is expected to focus on cheese production, leading to a growth of 1.7% p.a.

In contrast to the trends in the Russian Federation, Turkey should experience a production expansion for meat. Herd size expansion and yield improvements will characterise beef, sheep and poultry production for Turkey, in part driven by a self-sufficiency policy for red meat over the outlook period. In parallel, Turkey's cotton sector, one of the highest-yielding globally, will also see production growth. Its output is based on non-GM seeds, and yield improvements will come from mechanisation, irrigation, and the use of improved seeds.

Oceania: Environmental regulations constrain growth of livestock sector

Oceania is an important agricultural producer and net exporter of meat, dairy and cereals. As is the case in most other regions, Oceania will expand production of its main commodities at a slower rate than in the last decade.

Despite productivity improvements foreseen over the coming decade, the global share of sheep meat from Australia and New Zealand will decline as developing countries expand output. This relative decline will occur in parallel to a slowdown in milk production from the region, emerging from land constraints and environmental restrictions. Consequently, milk output in New Zealand will expand at 1.5% p.a., down from 3.3% p.a. over the last decade. The region is also an important producer of skim milk powder (SMP) and whole milk powder (WMP); by 2027, it will account for 17% of SMP and 27% of WMP global supplies.

Coconut oil production will become the focus of niche production by countries in the region over the coming decade, yielding per annum growth in vegetable oil production of 2.2%. For cotton, an increase of 16% in area harvested over the outlook period will underpin output growth in the region. In Australia, cotton production is expected to expand by 23%, due in part to the adoption of GM varieties.

Total fish production will increase by 19%, and will continue to play a major role in the food security of many Small Island Developing States in the region.

Western Europe: High productivity maintained within tight regulatory and resource base

The countries of Western Europe (which includes the European Union, Switzerland, and Norway) hold significant shares in the global production of other coarse grains (barley, oats, rye; 31% of global production); other oilseeds (rapeseed, sunflower; 20%); wheat (20%); milk (21%); and meat (15%). The coming decade will see declines in these global shares as other countries and regions report faster growth.

The decline will be especially pronounced for biodiesel, where the region's share of global production will fall from 40% to 34%, as output falls by around 4% over the outlook period, in line with lower demand for diesel. Despite this diminishing share in global production, Western Europe will remain the world's second largest producer of biodiesel. A major uncertainty in the region is the potential reduction in obligatory blending which, if implemented, would drastically reduce production.

Total agricultural and fish production for the region will grow by about 3% by 2027, making it the region with the slowest expansion of production for the projection period. Despite this slow growth, and the region's limited potential for area expansion, its strength as a region of high productivity and continued high yields allows it to remain a major global provider of numerous agricultural commodities.

As the area harvested for various crops such as other oilseeds, sugar beet, and roots and tubers is expected to contract over the outlook period, crop production growth will come predominantly through yield improvements, which is notable for a region that already reports some of the highest yields in the world across commodities. Fish production will also show limited growth, mainly due to strict management and environmental policies.

The EU system of sugar quota was abolished in 2017. In the past, the quota system kept EU sugar prices above the world market price while limiting producers' ability to respond to these higher prices. The anticipated end of the quota system led to a 14% increase in sugar beet area in 2017 compared to the previous year, but in the coming decade, as EU prices fall in line with global markets, the area dedicated to sugar beet is expected to contract again to pre-2017 levels. At the same time, sugar beet yields will continue to grow. The net result is that sugar beet output in the European Union will expand by 2.5% between the base period (2015-17) and 2027.

Strict management and environmental policies will constrain growth in fish, livestock and dairy production. Policies such as the EU nitrate directive, which imposes constraints on nitrates from agricultural production to protect water quality, is expected to constrain growth in milk output, and in turn beef production, over the outlook period. Despite the slowdown in fresh milk production, which will grow by 8% over the next decade (compared to 10% for the past ten years), SMP and WMP output for the region will expand by 10% and 18% by 2027. For WMP, this growth is substantially above that of the previous decade.

Trade

Specialisation between regions is increasing

Differences in climate and geography, including the availability of good agricultural land, determine the pattern of comparative advantage in producing different agricultural commodities. Together with differences in population density and population growth, as well as policy factors, this determines trade flows between regions. Countries with slow population growth, low population density and favourable natural endowments tend to become exporters of agricultural commodities, while countries with rapid population growth, greater population density, and less favourable natural endowments tend to become importers.

Figure 1.18 shows the historic and projected evolution of agricultural trade balances by region. These balances broadly reflect the forces described above and are projected to become more accentuated over time in most regions.

Figure 1.18. Agricultural trade balances by region, in constant value, 1990-2027

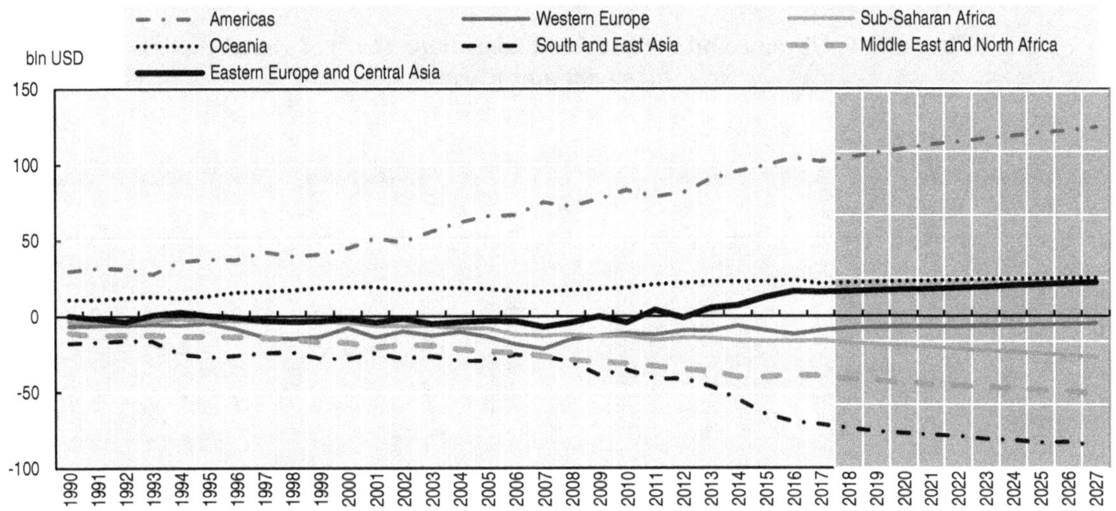

Note: Net trade (exports minus imports) of commodities covered in the *Agricultural Outlook*, measured at constant 2004-06 USD. Regions are as defined in the Production section.

Source: OECD/FAO (2018), "OECD-FAO Agricultural Outlook", OECD Agriculture statistics (database), http://dx.doi.org/10.1787/agr-outl-data-en.

StatLink http://dx.doi.org/10.1787/888933742226

Net exporters: Traditional suppliers expand market shares for most commodities

The Americas and Oceania are traditionally net exporters of agricultural commodities. The overall surplus for the Americas is split roughly equally between North America (United States and Canada) and Latin America and the Caribbean (most notably Brazil and Argentina). In Oceania, Australia accounts for roughly 60% of the overall surplus, with New Zealand accounting for the remainder.

While Oceania's agricultural trade surplus has remained stable over time, there has been a strong growth in the agricultural trade surplus of the Americas. Net exports have increased over time as producers responded to higher international demand for maize, soybean, and meat, among other commodities. The positive trade balance of the Americas is expected to expand further over the projection period.

In recent years, Eastern Europe and Central Asia has emerged as an important agricultural exporter. This change can be traced to improved export performance of the Russian Federation and Ukraine. The Russian Federation has switched from being a net importer to net exporter since around 2013. Agricultural trade in Ukraine was roughly balanced until 2007, when a strong growth in net exports began. The strong growth in Russian and Ukrainian exports is reflected in these countries' shares of global maize and wheat exports (Figure 1.19). Before 2008, Ukraine accounted for less than 5% of global maize exports. By 2011, this share had grown to 15%. The Russian Federation's share of global maize exports remains more modest, but nonetheless has grown from practically 0% in 2010 to 4% of the global total. In wheat, the relative positions are reversed. Both Ukraine and the Russian Federation have a history of exporting wheat surpluses globally, although export shares tended to be highly variable before 2012. Since then, the export shares have grown while becoming less volatile. Ukraine now accounts for 9% of global wheat exports, while the Russian Federation is now the largest exporter, providing 19-20% of global exports.

Figure 1.19. Ukraine and the Russian Federation: Share of global exports for maize and wheat

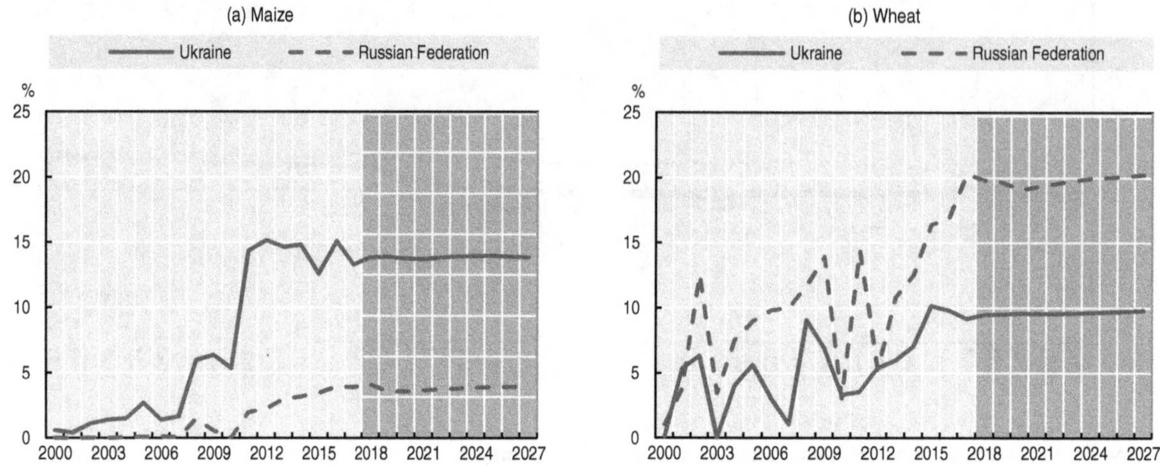

Source: OECD/FAO (2018), "OECD-FAO Agricultural Outlook", OECD Agriculture statistics (database), http://dx.doi.org/10.1787/agr-outl-data-en.

StatLink http://dx.doi.org/10.1787/888933742245

Net importers: Rising trade deficits among countries with rapid population growth

South and East Asia is a major net importer, although aggregate figures for the region hide considerable heterogeneity across countries. While Indonesia and Malaysia are established as net exporters (in large part due to palm oil), Japan is historically a net importer, although its agricultural trade deficit has been broadly constant over time. Conversely, since 2000 the Chinese agricultural trade deficit has grown strongly, accounting in 2017 for USD 40 billion of the region's USD 70 billion trade deficit (in constant 2004-06 dollars). Net imports in China (and hence in South and East Asia as a whole) are expected to grow in the coming decade, but at a slower rate.

Two regions which have similarly experienced increasing agricultural trade deficits are the Middle East and North Africa, and Sub-Saharan Africa. However, the role of imports in satisfying the consumption needs of these two regions varies considerably. While imports represent nearly 20% of consumption of major food commodities in Sub-Saharan Africa, approximately 57% of consumption is met through imports in the Middle East and North Africa region. The evolution of import dependence in the Middle East and North Africa is discussed in more detail in Chapter 2.

Western Europe's agricultural trade deficit (mostly attributable to the European Union) peaked in 2007. Since then, the deficit has fallen by around half to some USD 10 billion (at 2004-06 prices), and is expected to further decrease by around half over the outlook period.

Trade in fish and seafood

Regional trends in the overall agricultural trade balance may mask differences in the pattern of net importers and net exporters of specific commodities, such as fish and seafood, one of the most intensively traded products covered in the *Outlook*. Whereas the United States is a large net exporter of agricultural commodities and China a large net importer, the situation is reversed for fish and seafood. Over time, such regional differences have become more pronounced; since the early 1990s, net imports have increased in the European Union, the United States and Sub-Saharan Africa (among others), while net exports have increased in Norway, Viet Nam and China. For net exports, Viet Nam and Norway are expected to continue increasing their exports but a decline is projected for China due to a reduction in its fish production coupled with growing domestic demand.

Agricultural trade growth is slowing

Across the commodities covered in the *Agricultural Outlook*, the growth of trade volumes is expected to slow down significantly, as shown in Figure 1.20. For some commodities such as skim milk powder, soybeans, and cereals, trade volumes grew strongly in the past decade with growth rates of 4% to 8% per year. In the coming decade, consistent with the slower growth of demand, trade volumes will grow at a much slower pace. The highest expected growth rate (for rice) is only 2.2% per year, while some commodities (e.g. biofuels) will barely register any trade growth at all.

Figure 1.20. Growth in trade volumes, by commodity

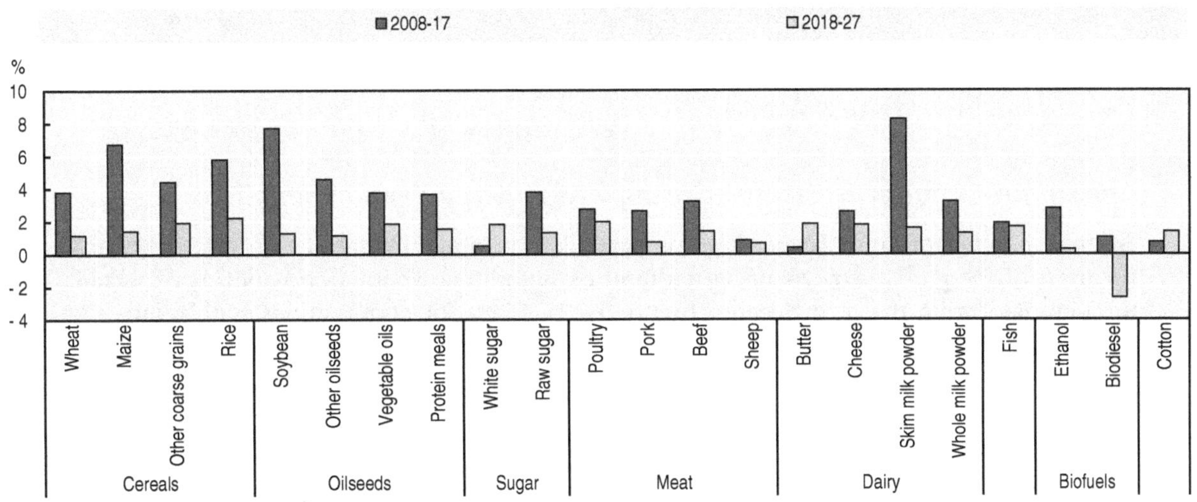

Note: Annual growth rate of trade volumes.

Source: OECD/FAO (2018), "OECD-FAO Agricultural Outlook", OECD Agriculture statistics (database), http://dx.doi.org/10.1787/agr-outl-data-en.

StatLink http://dx.doi.org/10.1787/888933742264

The importance of trade differs by commodity, as shown in Figure 1.21. For many agricultural commodities the share of production traded is low. Less than 7% of global pork production is traded internationally, and only around 8% of global butter production; the share is 9% for rice and 10% for biodiesel. Only for some commodities does trade represent at least one-third of global production. This is the case for cotton, sugar and soybeans, and also for vegetable oils and milk powders, which have a higher degree of processing.

Milk powders in particular are used as a low-cost way of shipping dairy, which explains their high trade share. As shown earlier, most dairy is consumed in the form of fresh dairy products (Figure 1.7), which are typically consumed domestically.

A low trade share at the global level does not mean that trade is not important. For many developing countries, imports of agricultural commodities are essential to guarantee food security. Import dependence is particularly high in the Middle East and North Africa, as discussed in Chapter 2.

Figure 1.21. Share of production traded

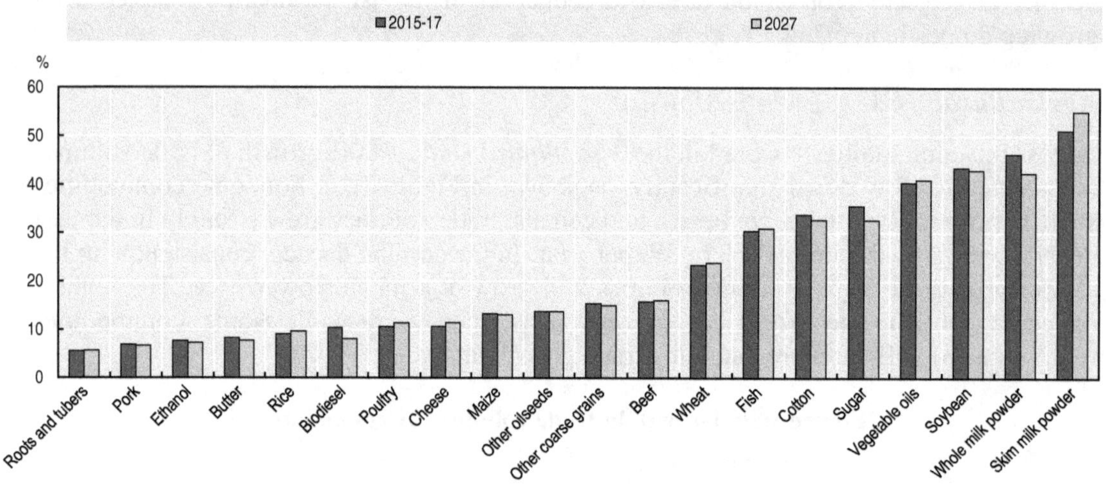

Source: OECD/FAO (2018), "OECD-FAO Agricultural Outlook", OECD Agriculture statistics (database), http://dx.doi.org/10.1787/agr-outl-data-en.

StatLink http://dx.doi.org/10.1787/888933742283

Agricultural exports to remain concentrated among a few key suppliers

A small number of countries with a comparative advantage in production often account for most of the global exports of agricultural commodities, a situation which is expected to continue during the next decade (Figure 1.22). Even for commodities with relatively less concentrated exports such as beef or wheat, the five leading exporters account for more than two-thirds of the global total. For soybean and pork, this share even exceeds 90%.

Moreover, even for some commodities where the share of the five main exporters is more modest, a single country often dominates. This is the case for sugar (where Brazil by itself accounts for 45% of global exports), other oilseeds (where Canada accounts for 54% of global exports), roots and tubers (where Thailand accounts for 56% of global exports), and several dairy products. For cheese, the European Union exports almost a third of the global total, a share which is expected to increase further. For butter and whole milk powder, New Zealand accounts for more than half of global exports.

Conversely, exports for skim milk powder are more equally distributed among key exporters. In 2015-17, the European Union, the United States and New Zealand had export shares of 30%, 25% and 19% respectively; in the coming decade, the United States is expected to increase its share of global exports but without changing this overall ranking. Exports are also less concentrated for fish for human consumption, where the top 5 exporters are expected to account for less than half of global export volumes by 2027.

Figure 1.22 also denotes for each commodity the value of the Hirschman-Herfindahl Index, a commonly used indicator of market concentration. Higher values of the Hirschman-Herfindahl index indicate greater concentration of exporters, whereas lower values are indicative of greater "evenness" with market shares more evenly distributed among participants. This measure conveys the relative dominance of exporters, complementing the information conveyed by the sum of market shares of the top 5 exporters. The Hirschman-Herfindahl Index will report relatively greater concentration when a single large exporter dominates the market, as is the case for sugar, other oilseeds, and whole milk powder.

Figure 1.22. Export shares of the top 5 exporters in 2027, by commodity

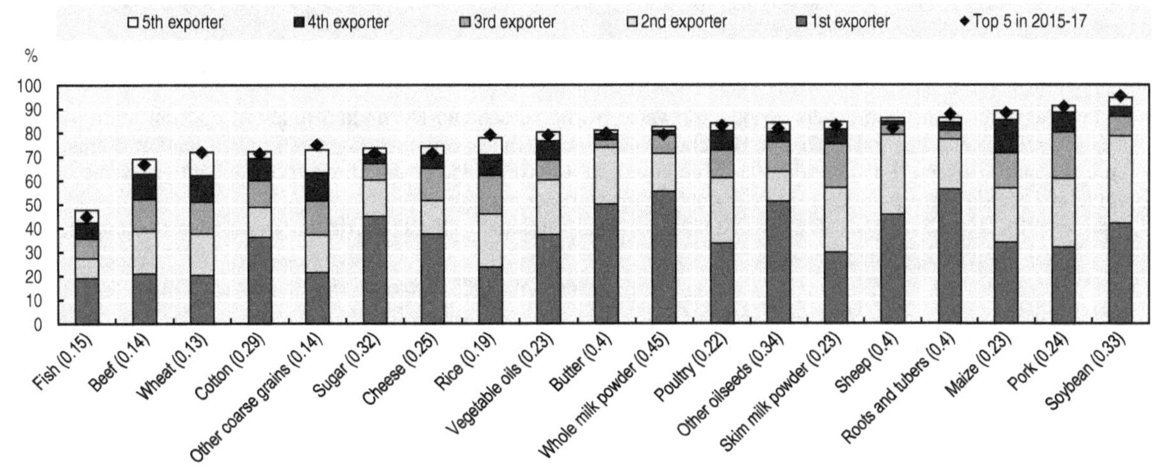

Note: The number in the brackets denotes the value of the Hirschman-Herfindahl Index of concentration of exports across countries for 2027. The Hirschman-Herfindahl Index equals the sum of squared market shares, here rescaled between 0 and 1, where a value closer to 0 corresponds to the absence of concentration and a value of 1 would correspond to a single country being the sole exporter.

Source: OECD/FAO (2018), "OECD-FAO Agricultural Outlook", OECD Agriculture statistics (database), http://dx.doi.org/10.1787/agr-outl-data-en.

StatLink http://dx.doi.org/10.1787/888933742302

Overall levels of concentration tend to be stable, and little change is expected over the coming decade. The high concentration of agricultural exports creates a risk of significant impacts on global markets if exports are interrupted, for instance because of adverse production shocks (e.g. poor harvests) or policy changes in the major exporting countries. Such interruptions could affect prices and local availability, with implications for food security.

Compared with exports, agricultural imports are typically more dispersed – that is, agricultural trade typically flows from a small number of exporters to a large number of importers (Figure 1.23). For rice and wheat, for instance, the five largest importers jointly account for less than 30% of global imports; for most commodities in the *Agricultural Outlook*, the share of the five largest importers is less than 60%. Similarly, the Hirschman-Herfindahl Index is generally lower for imports than for exports.

Figure 1.23. Import shares of top 5 importers in 2027, by commodity

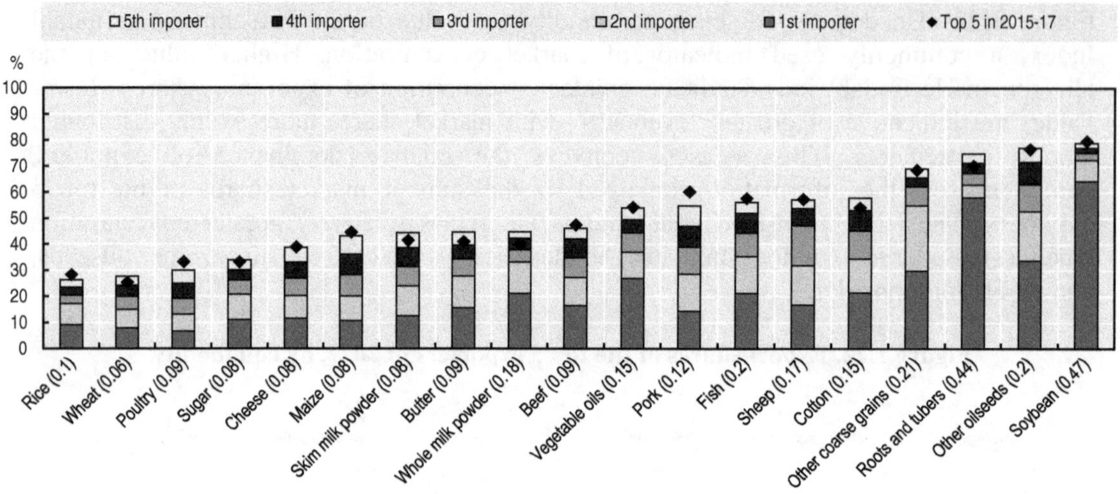

Note: The number in the brackets denotes the value of the Hirschman-Herfindahl Index of concentration of imports across countries for 2027. The Hirschman-Herfindahl Index equals the sum of squared market shares, here rescaled between 0 and 1, where a value closer to 0 corresponds to the absence of concentration and a value of 1 would correspond to a single country being the sole importer.

Source: OECD/FAO (2018), "OECD-FAO Agricultural Outlook", OECD Agriculture statistics (database), http://dx.doi.org/10.1787/agr-outl-data-en.

StatLink http://dx.doi.org/10.1787/888933742321

Notable exceptions are oilseeds (soybeans and other oilseeds), roots and tubers, and other coarse grains, where Chinese demand dominates. China currently is responsible for 63% of all global soybean imports, a share which is expected to increase somewhat in the coming decade. For roots and tubers, China is projected to increase its share of global imports from 53% to 58%. Soybeans and roots and tubers also have a high level of exporter concentration. Global soybean trade is hence dominated by US and Brazilian exports to China, while global trade in roots and tubers (cassava) is dominated by Thai and Vietnamese exports to China.

As with exports, the degree of import concentration by product will change over the next decade but without a clear trend towards higher or lower concentration. Greater import concentration is projected for skim milk powder, cotton and roots and tubers, among other commodities; while greater import dispersion is projected for poultry, beef and especially pork. For pork, while global trade is expected to continue growing, the volumes imported by the two main importers (China and Japan) are expected to decline over the outlook period. China is projected to fall below Japan as largest pork importer; together, these countries are expected to account for 29% of global imports in 2027 as opposed to 34% in the baseline period.

Prices

Real prices for most commodities are expected to fall

The *Outlook* uses recorded prices at main markets (e.g. US gulf ports, Bangkok) of each commodity as international reference prices and provides projections for these prices. Near-term price projections are still influenced by the effects of recent market events (e.g. droughts, policy changes), whereas in the outer years of the projection period, they are driven by fundamental supply and demand conditions.

Prices of different commodity groups such as cereals, oilseeds, dairy, and meat are highly correlated. In the coming decade, prices for these key commodity groups are projected to fall in real terms (Figure 1.24). This implies real prices are expected to be below the peaks seen in 2006-8 for cereals and oilseeds and in 2013-14 for meat and dairy, yet above the levels of the early 2000s.

Figure 1.24. Medium-term evolution of commodity prices, in real terms

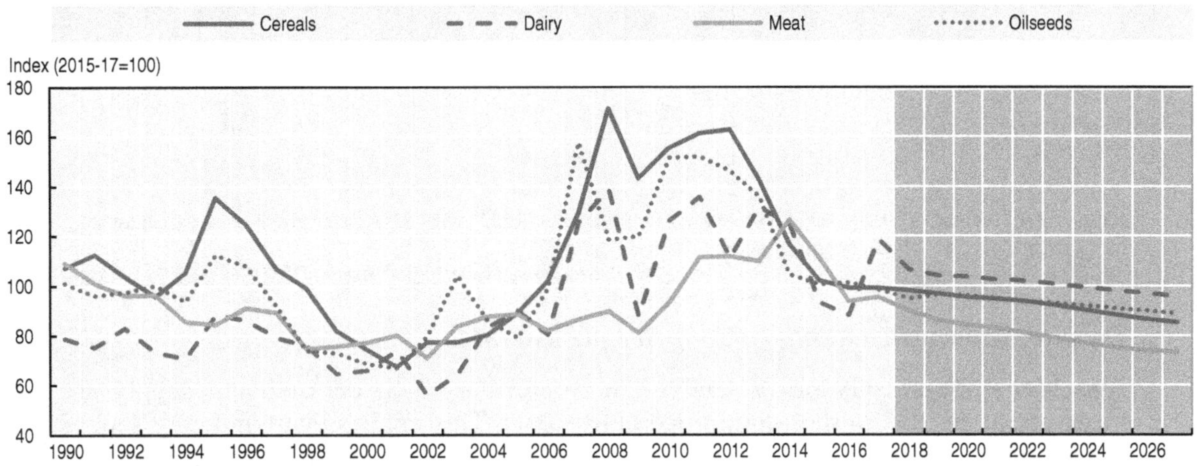

Source: OECD/FAO (2018), "OECD-FAO Agricultural Outlook", OECD Agriculture statistics (database), http://dx.doi.org/10.1787/agr-outl-data-en.

StatLink http://dx.doi.org/10.1787/888933742340

A more detailed view per commodity is provided in Figure 1.25 which shows the projected average annual real price change over the outlook period. Among generally declining real prices, projected trends for dairy stand out. Following the "butter bubble" in 2017, real prices for butter are expected to exhibit an average annual decline of 2% per year as prices come down further at the beginning of the projection period, but prices of skim milk powder are expected to increase by 1% per year. Along with whole milk powder, it is one of the only commodities covered in the *Agricultural Outlook* where prices are not projected to fall in real terms.

Trends in real prices for agricultural commodities reflect a balance of factors which would lead to higher prices (such as higher demand induced by population growth and higher incomes) and factors which would tend to reduce prices (such as productivity improvements, which allow greater output without using additional inputs). The pattern of real price decreases shown in Figure 1.25 indicates that under the assumptions made in the *Agricultural Outlook*, price-reducing factors, principally productivity growth, are expected to dominate in the coming decade.

Figure 1.25. Average annual real price change for agricultural commodities, 2018-27

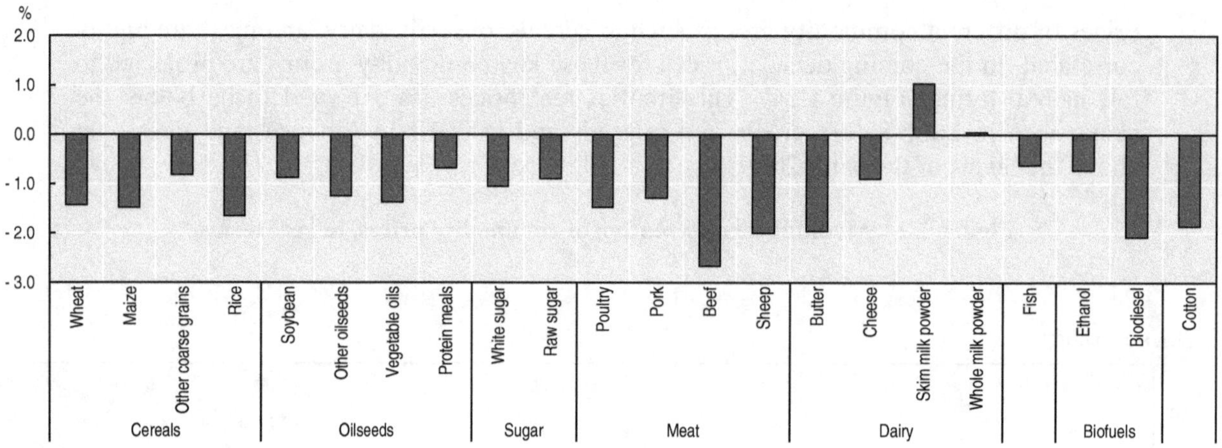

Source: OECD/FAO (2018), "OECD-FAO Agricultural Outlook", OECD Agriculture statistics (database), http://dx.doi.org/10.1787/agr-outl-data-en.

StatLink http://dx.doi.org/10.1787/888933742359

Despite a downward trend, risk of price peaks remains

Agricultural commodity prices tend to be volatile, as both demand and supply are relatively insensitive to short-term price movements. This implies that temporary shocks or uncertainty in the projections will have a relatively greater impact on prices than on consumption or production levels. The price trends presented here summarise the interplay of fundamental supply and demand factors, but short-term volatility may lead to considerable deviations from the trend.

To assess the uncertainty around prices, a partial stochastic analysis was performed on the projections of the *Agricultural Outlook.* The stochastic analysis simulates the potential variability of agricultural markets using 1 000 different scenarios for macroeconomic and other variables, such as oil prices, economic growth, exchange rates and yield shocks. In each scenario, the Aglink-Cosimo model underlying the *Outlook* projects different outcomes for prices. These can be used to provide an indication of the sensitivity of the estimates in the *Outlook.*

The degree of variation included in the stochastic analysis is based on the historical variability, which means more extreme shocks than those observed in the past are not incorporated in the stochastic analysis. Moreover, the analysis is only partial in that it does not capture all the sources of variability that can affect agricultural markets. For example, uncertainty related to animal diseases is not captured, as this factor is hard to quantify. The major sources of uncertainty in agricultural markets included in the stochastic analysis are (Araujo-Enciso et al, 2017):

- *Global macroeconomic drivers:* Values of 32 variables including the real Gross Domestic Product (GDP), the Consumer Price Index (CPI) and the GDP Deflator in the United States, the European Union, China, Japan, Brazil, India, the Russian Federation and Canada; national currency-US dollar exchange rates for these regions; and the world crude oil price.

Figure 1.26. Evolution of real prices for selected commodities

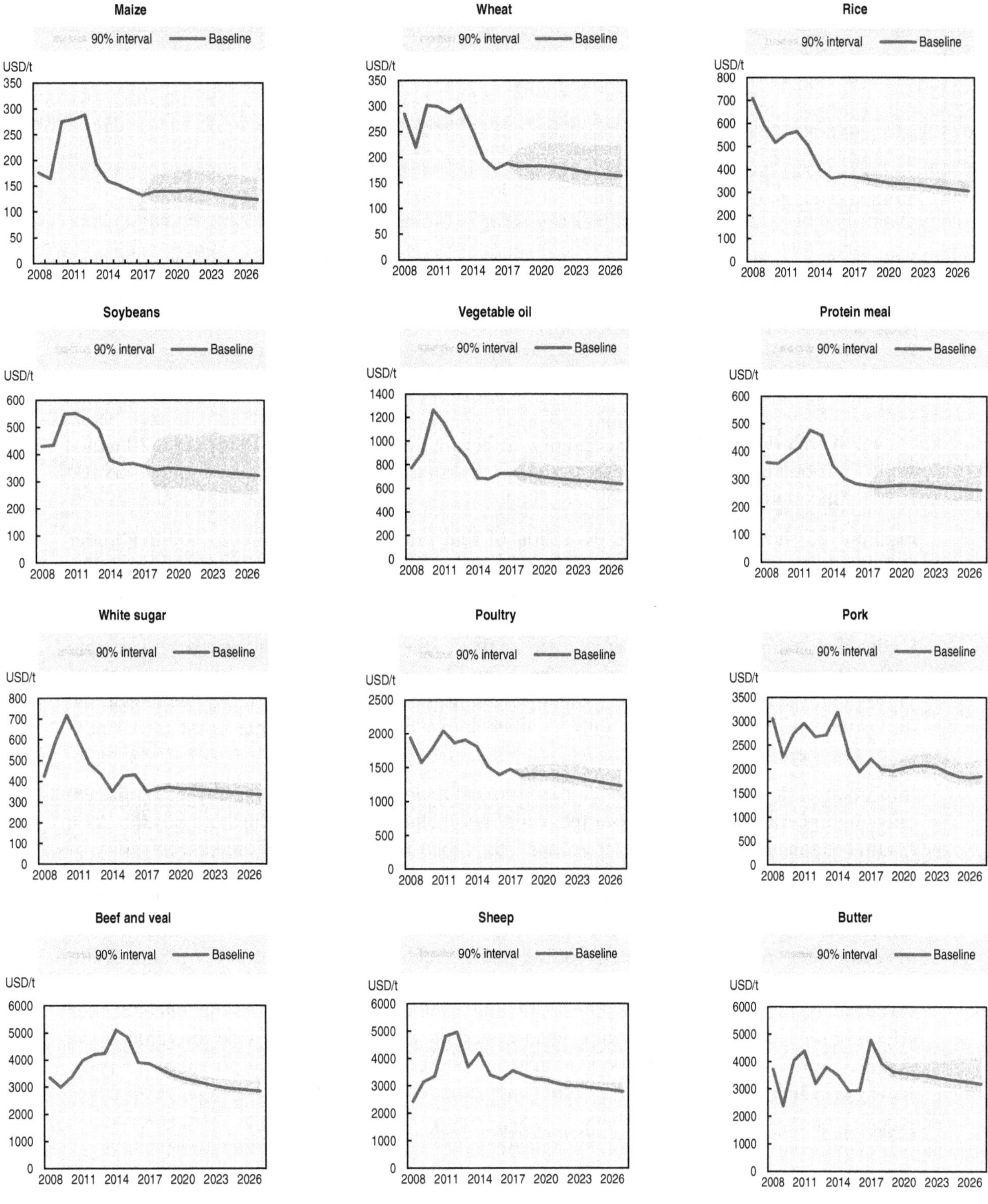

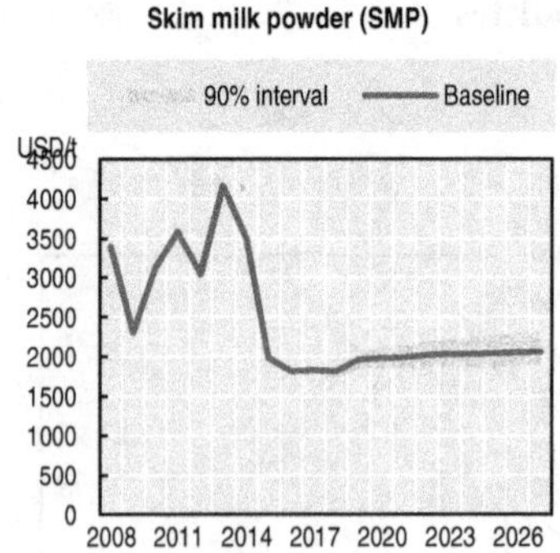

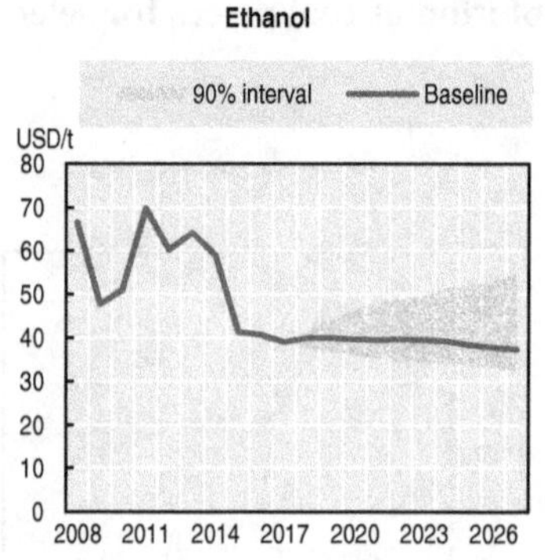

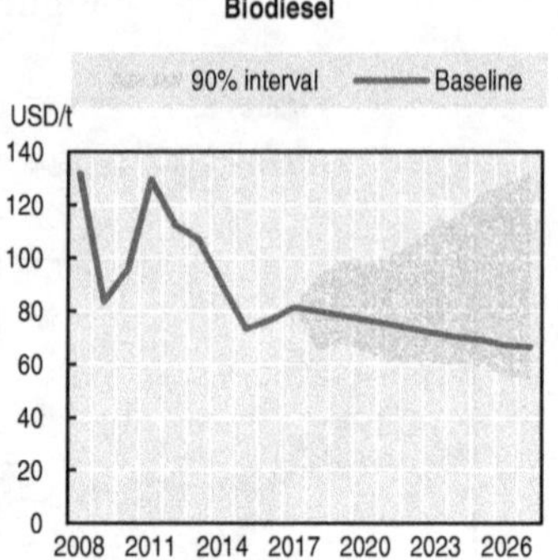

Note: Charts show the evolution of real prices over the outlook period in the baseline projection (solid blue line) as well as the 90% interval from a partial stochastic analysis (see text for details). Cotton and fish are not included in the stochastic analysis and are hence not shown here. Real prices are defined as nominal world prices deflated by the US GDP deflator (2010=1).
Source: OECD/FAO (2018), "OECD-FAO Agricultural Outlook", OECD Agriculture statistics (database), http://dx.doi.org/10.1787/agr-outl-data-en.

StatLink http://dx.doi.org/10.1787/888933742378

- *Agricultural yields*: Uncertainty affecting the yields of 17 crops in 20 major producing countries is also analysed, giving a total of 78 product-country-specific uncertain yields.

Figure 1.26 shows the expected evolution of real prices for selected commodities under the baseline scenario of the *Agricultural Outlook* as a solid line in each chart. The sensitivity of the price projections is shown as a 90% confidence interval around the projection; 90% of simulated prices in the stochastic analysis fall in this grey range. Under the assumptions of the stochastic analysis, the likelihood that prices will remain within the range is 90% in any given year. The likelihood that prices remain in this range throughout the decade is therefore much lower, at $(0.90)^{10}$ or around 35%. The likelihood that prices will fall outside the range (either above or below) at some point in the next decade is therefore 65%.

Importantly, this grey range does not capture all uncertainty around the projected prices, only the uncertainty coming from the variables included in the stochastic analysis. As a result, the range tends to be larger around crops than livestock, given the susceptibility of yields to weather conditions. Among crops, the price of rice varies least across the different simulations of the stochastic analysis, in part because paddy rice yields are less sensitive to weather conditions once planting decisions have been made. (Weather shocks instead affect area planted, since flooding paddy fields is a precondition for planting, but such area variations are not currently included in the stochastic analysis). By contrast, the variation is highest for biofuels (ethanol and biodiesel), which combine uncertainties affecting physical production with greater uncertainty on the demand side. In general, the degree of uncertainty tends to be asymmetric, as there is more upside potential for price spikes than for price declines.

Projected evolution of the FAO Food Price Index

Another way of assessing the evolution of prices is through the expected future path of the FAO Food Price Index. This index, introduced in 1996, captures the development of nominal prices for a range of agricultural commodities in five commodity groups, weighted with the average export shares of these groups in 2002-2004. As this commodity price index is similar in commodity coverage to the *Agricultural Outlook*, it is possible to project the future evolution of the FPI as a summary measure of the evolution of nominal agricultural commodity prices (Figure 1.27).

Figure 1.27. Projected evolution of the FAO Food Price Index

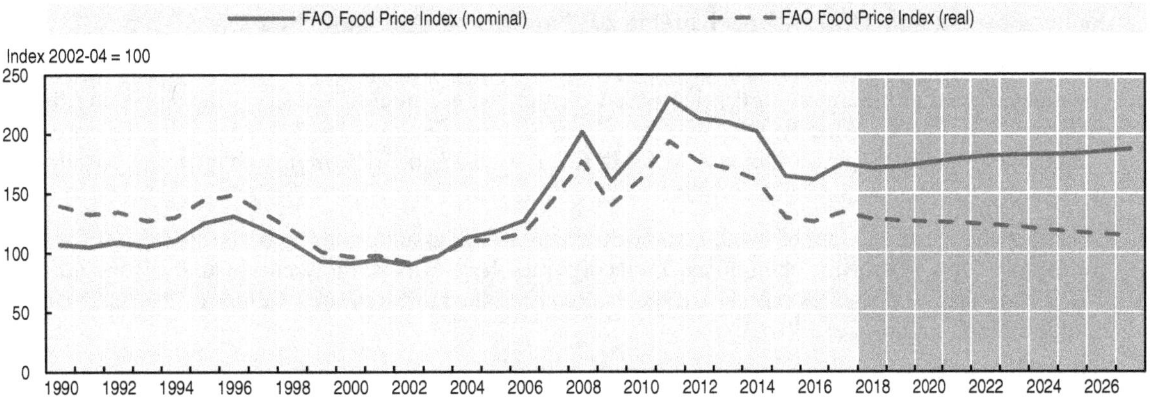

Note: Historical data is based on the FAO Food Price Index, which collects information on nominal agricultural commodity prices; these are projected forward using the Agricultural Outlook baseline. Real values are obtained by dividing the FAO Food Price Index by the US GDP deflator (2002-04 = 1).
Source: FAO World Food Situation (http://www.fao.org/worldfoodsituation/foodpricesindex/en/) and OECD/FAO (2018), "OECD-FAO Agricultural Outlook", OECD Agriculture statistics (database), http://dx.doi.org/10.1787/agr-outl-data-en.

StatLink http://dx.doi.org/10.1787/888933742397

Based on the supply and demand conditions projected in the *Outlook*, nominal agricultural commodity prices as summarised by the FAO Food Price Index are expected to grow by only 0.7% per year over the coming decade. In real terms, the FAO Food Price Index is expected to decline over the coming decade. Both nominal and real prices are expected to remain below the peaks reached between 2008 and 2014, but above the levels seen in the early 2000s.

Risks and uncertainties

The *Agricultural Outlook* combines projections using the Aglink-Cosimo model with expert judgment on the likely evolution of drivers of agricultural markets. The projections in the *Outlook* are therefore sensitive to the underlying assumptions, such as the assumptions on macro-economic conditions and relevant policies discussed in Box 1.6. While based on the best available information at the time of preparation, these assumptions are inherently uncertain. Moreover, a number of factors not explicitly taken into account could affect global agricultural markets in the coming decade. Uncertainty on such issues tends to accumulate over time. Over the ten-year horizon of the *Outlook*, temporary deviations from the trend may hide the actual trend, even if the projections of the *Outlook* are sound.

Some of the uncertainties can be quantified. For instance, the impact of a different oil price scenario is explored in Box 1.3. Moreover, the partial stochastic analysis introduced in the previous section can provide useful information on the sensitivity of the projections in the *Outlook* to changes in global macroeconomic conditions and agricultural yields. Finally, several other factors are harder to quantify; their potential impact is discussed below.

Box 1.3. The impact of an alternative oil price scenario

The assumption on crude oil prices during the projection period is based on the average crude oil price projected by the World Bank Commodities Price forecasts, as released in October 2017. These projections imply that nominal oil prices increase at an average annual rate of 1.8% over the outlook period, from USD 54.7 per barrel in 2017 to USD 76.1 per barrel by 2027.

To test the sensitivity of the *Outlook* projections to this assumption, a scenario analysis was conducted with an alternative oil price based on the "New Policies Scenario" developed by the International Energy Agency (IEA) in its *World Energy Outlook 2017*. Under this alternative scenario, nominal oil prices increase to USD 122.2 in 2027, or 61% higher than in the baseline scenario.

A large change in oil prices would also affect the GDP assumptions underlying the *Outlook*, especially for oil-exporting economies. To incorporate these effects, the scenario analysis included GDP responses to oil prices based on a recent study by the Joint Research Centre of the European Commission (Kitous et al., 2016).

Higher oil prices increase agricultural production costs through higher prices for fuel and fertiliser, as well as through general cost increases induced by higher inflation. Higher fuel prices can also affect demand for agricultural commodities through biofuels markets, through two opposing effects. On the one hand, higher prices depress the demand for transportation fuels, which in turn reduces the demand for biofuels that is due to mandatory blending. On the other hand, a higher crude oil price leads to a substitution in favour of biofuels. This effect would be more pronounced for biodiesel than for ethanol, for which the share in gasoline fuel is already close to its technical maximum in several main markets.

The scenario suggests that higher crude oil prices would have a negative but small impact on the production of most agricultural commodities. For instance, for maize, global production would be 0.7% lower than under the baseline projections. Stronger effects are found for biofuels. Higher oil prices would stimulate the global production of biodiesel by 2.5% compared to the baseline while global ethanol production would decrease by 1.5%.

Higher crude oil prices would also affect agricultural prices. Nominal prices of maize, wheat, soybeans and vegetable oil would all be 10-11% higher than under the baseline while nominal livestock and dairy prices would be 6-8% higher. A stronger price increase is expected for biodiesel, where higher demand, production costs and inflation lead to nominal prices that are 27% above those in the baseline.

Several factors affect the "pass-through" of oil prices to agricultural commodity prices. The scenario assumes that the higher oil price is caused by supply-side factors, so that higher oil prices reduce demand for transportation fuel, in turn reducing the demand for biofuels due to mandatory blending. If higher oil prices were caused by an increased demand for transportation fuel, it would be accompanied by a stronger growth in biofuels demand and hence a stronger increase in agricultural prices.

A second factor is the impact of higher oil prices on fertiliser prices. Traditionally, high oil prices led to high prices of natural gas, a main input for nitrogen-based fertilisers. The price of natural gas was in the past often indexed to crude oil prices, creating a direct link. In recent years, however, natural gas prices have shown signs of becoming "decoupled" from oil prices. This would weaken the link between oil and fertiliser prices. On the other hand, it seems likely that a large increase in crude oil prices sustained over a decade, as considered in the scenario, would be accompanied by higher natural gas prices – whether due to the way in which natural gas is priced, or due to substitution effects. The scenario therefore assumes that crude oil prices will indeed affect fertiliser prices.

Source: Kitous, A., et al. (2016) Impact of low oil prices on oil exporting countries, *JRC Science for Policy Report*, EUR 27909 EN (doi:10.2791/718384).

Partial stochastic analysis

In the previous section, a partial stochastic analysis was used to provide an indication of the range of uncertainty around projected real prices for various commodities. The stochastic analysis also offers insights on other aspects of the *Outlook*. One way of representing and comparing the impact of uncertainty on projected outcomes is the coefficient of variation in the last projection year, 2027. The coefficient of variation (CV) is defined as the standard deviation divided by the mean, and can therefore be interpreted as a percentage deviation from the "central" projection in the *Outlook*.

Figure 1.28 compares the coefficients of variation of global consumption, production, trade, (nominal) prices and stocks of maize. Whereas the coefficient of variation for global consumption is around 1%, variability of production is larger at almost 3%. For trade, the coefficient of variation is around 5%. Variability of prices is much larger at 11%, while the highest variability is for stocks, at 16%.

This result captures two essential characteristics of global agricultural markets. First, the demand and supply of many agricultural commodities are relatively less sensitive to prices. Shocks to demand or supply therefore lead to relatively large adjustments in prices. Second, trade and stocks serve as buffers and are therefore more variable. Stocks can be used to smooth consumption in the face of fluctuations in production. Likewise, trade allows countries to increase imports to keep consumption more stable in years where production is low.

Figure 1.28. Maize: Coefficient of variation in 2027

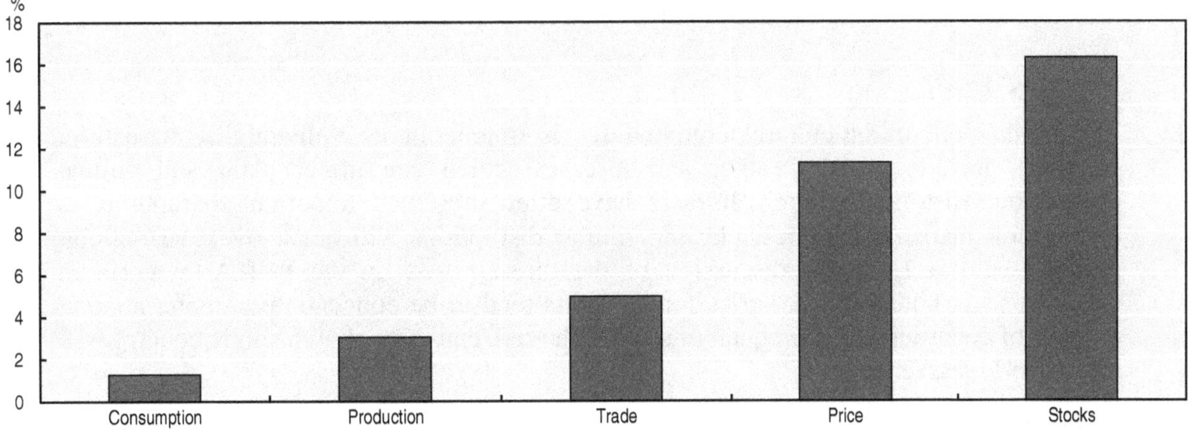

Source: OECD/FAO (2018), "OECD-FAO Agricultural Outlook", OECD Agriculture statistics (database), http://dx.doi.org/10.1787/agr-outl-data-en.

StatLink http://dx.doi.org/10.1787/888933742416

Other uncertainties affecting the Outlook

While the partial stochastic analysis captures uncertainty around a range of factors affecting the evolution of agricultural markets, many other uncertainties are harder to quantify but no less important, in particular those associated with government policies.

Demand

On the demand side, an important source of uncertainty relates to biofuel policies in major markets, notably China. The Chinese government recently proposed a new nationwide ethanol mandate, which would expand an earlier mandate that had been in force in 11 trial provinces to the entire country by 2020. The potential implications are discussed in more detail in the Biofuels chapter, but preliminary estimates show that the policy would increase Chinese ethanol use by 18 bln L to 29 bln L. To put this increase in context, world ethanol production in 2027 is currently projected at 131 bln L. If inputs are sourced domestically, Chinese maize reserves could be used in large part; but if the additional demand is instead satisfied by imports, the effect on agricultural markets could be large.

Changing consumer preferences could also impact markets. Some evolutions in consumer demand can be projected from current trends, such as the decreasing role of cereals and the increasing demand for proteins as average incomes grow. Other changes, such as a rise in vegetarian or vegan lifestyles or an increasing preference for local or organic food, are harder to assess, but these tend to be relatively slow-moving trends and are often of limited importance to global markets. Food health scares, by contrast, have the potential of reducing consumer demand in the short run, sometimes with lasting consequences.

Obesity and overweight are increasingly recognised as public health problems in a large number of countries. Various policies have been introduced to stem the rise of obesity, varying from the provision of information and education to labelling and product formulation requirements, and subsidies and taxes (most notably on sugar and sweetened beverages). Further measures may be deployed over the projection period, with a view to affecting the levels of calorie consumption as well as the composition of diets.

Supply

The production of agricultural commodities is unique in its vulnerability to natural elements, including bad weather and diseases which can affect plant and animal production. Historically, such diseases have often presented important disruptions to agricultural markets; it is possible that similar disruptions will occur over the outlook period (see Box 1.4 for a discussion of the threat posed by the Fall Armyworm in particular). As noted above, agricultural exports tend to be concentrated among a small number of countries; all else equal, this raises the risk that a shock in a single country will affect world markets.

Regulatory changes can impact agricultural production, for instance through the introduction of measures that ban or raise the costs of certain production practices (e.g. the use of neonicotinoid pesticides). Similarly, policies to mitigate climate change could impact agriculture, in particular ruminant production, which contributes to methane emissions. On the other hand, developments of new technologies such as digital and precision agriculture or new plant breeding techniques could improve agricultural productivity beyond the rate currently projected in the *Outlook*.

Agricultural input industries are currently witnessing a trend of increasing consolidation and market concentration. Such trends have been seen in crop protection chemicals, seeds and biotech, and fertiliser markets, among others, raising concerns that less competition may reduce private spending on research and development (R&D).

For the fisheries and aquacultures sector, an important source of uncertainty relates to changes in policies being implemented by China, which can have an impact on global supply, demand, and prices due to the key role played by China in the sector. The potential implications are discussed in more detail in the fish and seafood chapter.

Box 1.4. Combatting the expanding Fall Armyworm infestation in Sub-Saharan Africa

Fall Armyworm (*Spodoptera frugiperda*) is an insect native to the Americas that was first detected in Central and Western Africa in early 2016. Since then, it has spread to most countries in Sub-Saharan African and is likely to reach North Africa (FAO, 2017). In the medium term, experts fear that infestations could extend into Southern Europe and Asia and, during the summer season, even reach Northern Europe. In the Americas, farmers, researchers and governments have been combatting Fall Armyworm for decades, keeping losses to a minimum. However, in Sub-Saharan Africa the majority of maize farmers are smallholders and do not have access to the necessary knowledge or inputs to fight this new pest. While some studies based on farmer's perceptions claim that in the absence of any control method Fall Armyworm can cause maize production losses up to 53% (Day et al., 2017), the majority of the field trials show yield losses below 20%.

The Fall Armyworm outbreak in Sub-Saharan Africa does not appear to have impeded the recovery in maize production following two consecutive years of severe drought conditions in Southern Africa. In 2017, cereal production increased by about 16 Mt compared to 2016, putting the aggregate output at 80 Mt, an above-average level. The *OECD-FAO Agricultural Outlook* foresees a continuation of this positive trend, with maize production projected to reach about 93 Mt in 2027 for the region. The projections assume Fall Armyworm control methods become effective enough to allow continued yield gains.

Nonetheless, such methods are not easily implemented and Fall Armyworm could become a threat to food security in the region. It holds the potential to endanger the production of cereals and other crops because, unlike the Americas, small-scale producers represent the vast majority of cereal production in Sub-Saharan Africa. Their crops are typically more vulnerable to pests and diseases, and their ability to cope with infestations is limited.

The *OECD-FAO Agricultural Outlook* projections take the Fall Armyworm into account as an important uncertainty. At the same time, severe production losses are expected to be prevented by initiatives already underway, notably FAO's five-year programme of "Sustainable Management of the Fall Armyworm in Africa." The programme incorporates the participation of researchers, governments, and small producers in Latin America with vast experience on managing Fall Armyworm. The methods and tools developed in Latin America are expected to prove effective in containing Fall Armyworm in Sub-Saharan Africa.

There is a possibility that the Fall Armyworm gradually moves to North Africa and from there to Europe and Asia. Unlike Sub-Saharan Africa, which is more of a regional market, the spread of Fall Armyworm to North Africa, Europe and Asia could pose problems for the global maize market, as those regions comprise major importers and exporters of maize. While it is still too early to assess the implications of such an outcome, efforts are already underway to ensure an effective monitoring and early detection of the pest. These efforts should eventually enable farmers and governments to take adequate and timely actions to contain the spread and mitigate the effects of Fall Armyworm.

Sources

FAO (2017), "Sustainable Management of the Fall Armyworm in Africa", FAO Programme for Action, 6 October, http://www.fao.org/3/a-bt417e.pdf

Day, R. et al. (2017), "Fall Armyworm: Impacts and Implications for Africa", *Outlooks on Pest Management*, Vol. 28, No. 5, pp. 196-201.

Trade

The international trade environment is facing increasing uncertainty in recent years, which may impact agricultural trade flows.

A number of current trade issues involving agricultural commodities (such as the Russian import ban, the dispute around Argentine and Indonesian biodiesel exports to the US, and China's anti-dumping probe into imports of US sorghum) may have important bilateral effects for specific commodities, but are not likely to have large effects at a global level and across different commodities (Box 1.5). However, even if such disputes are eventually resolved, they may end up permanently changing trade flows between countries, as exporters find new markets and importers find new sources of supply.

Brexit – the announced exit of the United Kingdom from the European Union – is currently still being negotiated; little is known about the exact arrangements that will govern agricultural policy in the United Kingdom and its trade relations with the European Union and other countries. While Brexit is likely to have a big impact on certain bilateral agricultural trade flows (most notably for beef, dairy and lamb), the effect on global agricultural trade is likely to be small.

In March 2018, eleven countries (Australia, Brunei, Canada, Chile, Japan, Malaysia, Mexico, New Zealand, Peru, Singapore and Viet Nam) signed the Comprehensive and Progressive Agreement for Trans-Pacific Partnership. Parties to the agreement are reducing tariffs on each other's agricultural imports, which is likely to intensify trading relations among participant countries. The agreement is also likely to have a negative effect on exports to countries that are party to the agreement by countries that are not. Again, these changes will impact individual countries and bilateral trade flows more than global agricultural markets.

The renegotiation of NAFTA, which is currently ongoing, could impact agriculture in North America. Agricultural trade has grown strongly because of NAFTA, leading to a highly integrated region. Currently, more than 25% of US maize exports go to Mexico, and one-third of US beef exports go to Canada and Mexico; disruptions to these trade flows could impact not only North American but global markets as well.

Box 1.5. Potential impacts of China imposing additional import tariffs on US agricultural products

China is the United States' largest trading partner. Total merchandise exports to the United States rose from USD 84 billion in 2000 to USD 506 billion in 2017. In terms of net trade, the United States has an annual deficit of about USD 375 billion in total merchandise trade, while it maintains a surplus of about USD 20 billion on agricultural products, with soybean exports accounting for USD 13 billion.

In March 2018, the United States imposed additional import duties on steel and aluminium products, and announced possible actions to respond to alleged unfair treatment of US companies seeking to do business in China on grounds of intellectual property infringement. Chinese authorities, in turn, suspended tariff concessions on multiple US products – including fruits, nuts and pigmeat – and announced eventual further duties on other agricultural products. Additional *ad valorem* duties of 25% have been put in place for pigmeat imports, and announced for soybeans and sorghum.

About 60% of US soybean exports are destined for China, which is highly dependent on imports for its domestic needs. In 2017, China imported an estimated 96 million tonnes, accounting for 64% of global soybean imports, while it produced around 13 million tonnes. Additional duties on soybeans would lower imports from the United States, but are likely to be offset by larger purchases from other suppliers, notably Brazil and Argentina. This could lead to a wider reallocation of trade, with US exports redirected to other markets, notably in Europe and Latin America, when the price spread between United States and Brazilian soybeans substantially widened. Indications of this have already been observed.

China has taken further measures to curb sorghum imports from the United States. In 2017, 80% of US sorghum exports were shipped to China, accounting for around USD 957 million. In February 2018, China self-initiated an antidumping and countervailing duty investigation on imports of United States sorghum, and therefore in principle outside the scope of retaliatory measures announced by Beijing. As of early April, China requests a provisional deposit on sorghum imports from the United States, equivalent to an *ad valorem* duty of 178.6%. This measure, applied to all US companies, has led to a halt of US exports and a redirection of vessels already underway to China. Higher trade barriers on China's sorghum imports could trigger secondary effects, potentially leading to a reduction in China's high maize stocks or stimulating the import of other feed grains, notably barley, which would open up opportunities for alternative suppliers.

China is the world's largest producer and importer of pigmeat. In 2017, it produced more than 53 million tonnes, about 45% of the global production, and imported an estimated 1.6 million tonnes. The industry relies heavily on soybean meal to feed pigs. Over the medium term, higher tariffs and hence higher costs for soybeans and feed grains would raise the costs of production for China's pigmeat industry. This, combined with the higher tariffs and hence higher prices for imported pork, could lead to noticeable increases in domestic pork prices. China may elect to source its needs from alternative suppliers such as the European Union, Canada and Brazil.

Across these major product categories, additional import tariffs would imply some immediate losses to both US suppliers and Chinese consumers. Beyond immediate dislocations the overall market effects should be modest as these are highly tradable products and China has the potential to source from other countries, while the United States has the potential to supply other markets. Nevertheless, diverting trade comes at a cost, especially due to the size of the United States - China soybean relationship and lack of alternative partners. The impact would be greater if China were to seek to meet the demand shortfall from domestic production.

Highlights of the commodity projections

Cereals

Global cereal production is projected to expand by 13% by 2027, accounted for in large part by higher yields. For maize and wheat, the Russian Federation is emerging as a major player on international markets, having surpassed the European Union in 2016 to become the top wheat exporter. For maize, market shares will increase for Brazil, Argentina and the Russian Federation while declining for the United States. Thailand, India, and Viet Nam are expected to remain the major suppliers on international rice markets, while Cambodia and Myanmar are projected to capture a greater share of the global export market. Over the projection period, prices are expected to increase slightly in nominal terms but decline modestly in real terms.

Oilseeds

Global oilseeds production is expected to expand at around 1.5% p.a., well below the growth rates of the last decade. Brazil and the United States will be the largest soybean producers, with similar volumes. Protein meal use will grow more slowly due to slower growth in livestock production and as the protein meal share in Chinese feed rations has reached a plateau. Demand for vegetable oil is expected to grow more slowly due to slower growth in per capita food use in developing countries and the projected stagnation in demand as feedstock for biodiesel. Vegetable oil exports will continue to be dominated by Indonesia and Malaysia, while soybean, other oilseeds and protein meal exports are dominated by the Americas. Prices are projected to increase slightly in nominal terms over the outlook period, with slight declines in real terms.

Sugar

Production of sugarcane and sugar beet are projected to expand at a slower pace than in the previous decade. Brazil is projected to remain the largest producer with strong growth prospects foreseen in India, China and Thailand. Demand for caloric sweeteners (sugar and high fructose corn syrup) is expected to grow faster than for most commodities. Per capita consumption is stagnant in developed countries and in some developing countries where consumption has reached levels that raise health concerns. In Asia and Africa, population growth and urbanisation are expected to sustain growth in sugar consumption. Brazil will continue to account for some 45% of global exports, making it the largest exporter. Sugar prices are projected to follow a moderate upward trend in nominal terms but a downward trend in real terms.

Meat

Global meat production is projected to be 15% higher in 2027 relative to the base period. Developing countries will account for 76% of the output growth, with poultry seeing the most rapid expansion. However, consumers in developing countries are expected to increase and diversify their meat consumption towards more expensive meat such as beef and sheepmeat. Import demand is expected to remain strong in Asia, with significant growth in the Philippines and Viet Nam; other main importers include China, Korea and Saudi Arabia. The combined export share of the two largest meat exporters, Brazil and the United States, is expected to increase to around 45%. Nominal meat prices are projected to gradually increase until 2027, while real prices are expected to trend downwards.

Dairy

Growth in world milk production is projected to increase by 22% over the projection period, with a large share of the increase coming from Pakistan and India. In 2027, these two countries are expected to jointly account for 32% of global milk production. Most of the additional production in these countries will be consumed domestically as fresh dairy products. Over the projection period, the European Union's share in global exports of dairy commodities is expected to increase from 27% to 29%. As the 2017 butter bubble continues to deflate, nominal and real prices for butter will decrease over the projection period. With the exception of milk powders, dairy prices are expected to decrease in real terms.

Fish

Global fish production will continue to grow, albeit at a much reduced pace compared with last decade. Additional output derives completely from continued but slowing growth in aquaculture, while capture fisheries production is expected to fall slightly. Policy changes in China imply a potentially large reduction in the growth of its aquaculture and capture fisheries output. Asian countries will account for 71% of the increase in fish consumption as food, and per capita fish consumption will increase in all continents except Africa. Fish and fishery products will continue to be highly traded; Asian countries will continue to be the main exporters of fish for human consumption while OECD countries will remain the main importers. Fish prices will all increase in nominal terms but remain broadly flat in real terms.

Biofuels

Given current policy developments and trends in diesel and gasoline demand, global ethanol production is expected to expand from 120 bln L in 2017 to 131 bln L by 2027, while global biodiesel production is projected to increase from 36 bln L in 2017 to 39 bln L by 2027. Advanced biofuels based on residues are not expected to take off over the projection period due to lack of investment in research and development. Trade in biofuels is projected to remain limited. Global biodiesel and ethanol prices are expected to decrease respectively by 14% and 8% in real terms over the next decade; however, the evolution of ethanol and biodiesel markets will continue to be shaped by policies and demand for transport fuel, which implies considerable uncertainty on these projections.

Cotton

World cotton production is expected to grow at a slower pace than consumption during the first few years of the outlook period, reflecting lower prices and releases of global stocks accumulated between 2010 and 2014. India will remain the world's largest country for cotton production, while the global area devoted to cotton is projected to recover slightly despite a decrease of 3% in China. Processing of raw cotton in China is expected to continue its long-term downward trend, while India will become the world's largest country for cotton mill consumption. In 2027, the United States remains the world's main exporter, accounting for 36% of global exports. Cotton prices are expected to be lower than in the base period (2015-17) in both real and nominal terms, as the world cotton price is continuously under pressure due to high stock levels and competition from synthetic fibres.

Box 1.6. Macroeconomic and policy assumptions

The main assumptions underlying the baseline projection

The *Outlook* presents a scenario that is considered plausible given assumptions on the macro-economic, policy and demographic environment, which underpins the projections for the evolution of demand and supply for agricultural and fish products. The macro-economic assumptions used in the *Agricultural Outlook* are based on the *OECD Economic Outlook* (November 2017) and the IMF's *World Economic Outlook* (October 2017). These and other assumptions are detailed in this box.

Global growth

After particularly weak growth in 2016, the upturn of the global recovery gathered strength in 2017, with growth of 3.6%. Similar rates of growth are expected in 2018 and 2019. In advanced economies, growth is accelerating in Europe, Canada, Japan and the United States, and inflation is still subdued, but growth at these rates may not be sustainable in the medium-term. World growth is mostly driven by emerging market and developing economies, but that growth remains uneven, in particular for some commodity exporters.

Growth in the United States is projected to increase to 2.2% in 2017 and 2.5% in 2018, boosted by fiscal stimulus, favourable financial conditions, and greater confidence among consumers and investors. Over the next ten years, growth is expected to be moderate at an average annual rate of 1.7%.

Figure 1.29. GDP growth rates in OECD and selected developing countries

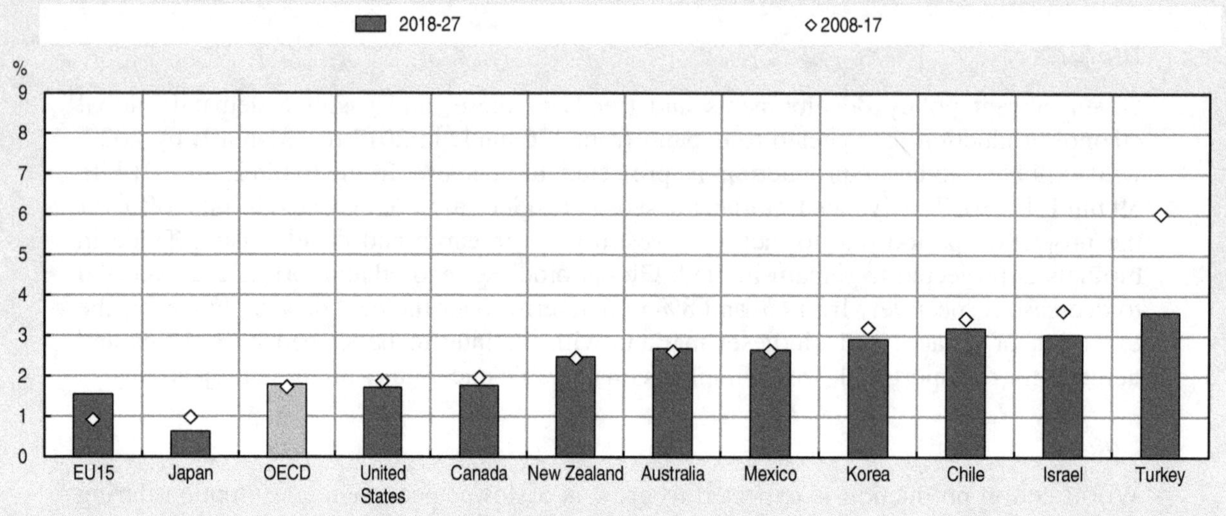

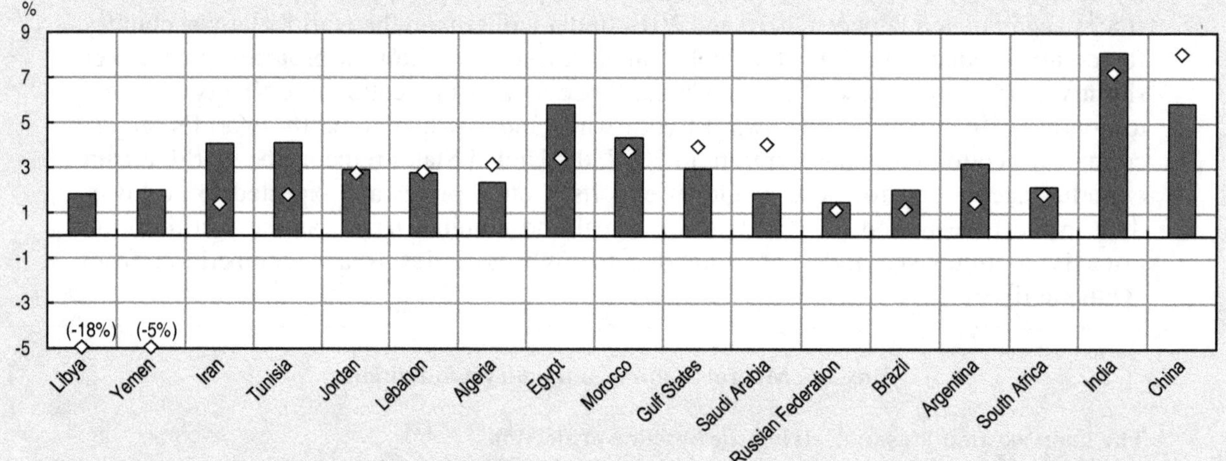

Note: Only selected developing countries shown in second panel. Assumptions for all countries are available in the online Statistical Appendix.

Source: OECD/FAO (2018), "OECD-FAO Agricultural Outlook", OECD Agriculture statistics (database), http://dx.doi.org/10.1787/agr-outl-data-en.

StatLink http://dx.doi.org/10.1787/888933742435

The recovery in the euro area is expected to gather strength this year, with growth at 2.1%, and be slightly down in 2019 at 1.9%, but is expected to remain moderate in the next decade due to weak productivity and low population growth. For EU15 members as a group, an annual average growth rate of 1.6% is expected during the projection period.

After a rebound of 1.5% in 2017, growth in Japan is projected to decrease again in 2018 and 2019, to 1.2% and 1.0%, respectively. Annual average GDP growth is expected to weaken further to 0.6% over the projection period due to a reduced labour force.

Among OECD countries, Turkey is expected to experience the highest growth rate over the next ten years, with an average annual rate of 3.6%, followed by Chile at 3.2%, Israel at 3.0%, Korea at

2.9%, and Australia and Mexico at 2.7%. Growth in Canada shows a strong recovery in 2017 at 3.0% but is projected to decrease in 2018 to 2.1% and not exceed 1.8% on average during the next decade.

Growth is projected to continue to slow down in China to an average of 5.8% over the next ten years compared to 8.0% during the last decade, while growth in India is anticipated to be strong at 8.1% p.a. on average.

After recessions in 2016, growth in Brazil, Argentina and the Russian Federation recovered in 2017 and is expected to average 2.0%, 3.2% and 1.5% p.a. respectively over the projection period. Growth in South Africa should average 2.2% over the ten-year period.

Economic growth in the Middle East and North Africa region is recovering following a recession induced by weak crude oil markets. Modestly stronger growth is projected through the medium term, with the region as a whole growing at an average of 3% p.a. over the outlook period, although growth is uneven among countries due largely to geopolitical factors. Egypt is anticipated to be the strongest growing country, with GDP increasing at 5.9% p.a. Other countries are projected to grow at between 2% and 5 % p.a., but some may not recover ground lost in the previous decade.

Emerging developing countries in Southeast Asia are projected to continue experiencing robust growth over the medium term, at least matching their performance of the previous decade. Growth in Viet Nam, Indonesia, and the Philippines is anticipated to be in the 5-7% p.a. range, while that in Thailand is at around 3.1% p.a.

In countries of the Latin America and Caribbean region, economic growth varies considerably by country. While Brazil and Argentina may grow relatively slowly in the next decade, other countries including Colombia and Chile are projected to average between 3% and 4% p.a.

Growth in developing and least developed African countries, while highly diverse, is projected to continue at higher rates in the next decade, and on a per capita basis may average over 3% p.a. Continued growth in most African countries will be contingent on firm commodity markets, and domestic policy reforms.

Population growth

World population growth is expected to slow to 1% p.a. over the next decade, compared to 1.3% in the last decade. Developing countries continue to drive this growth, particularly Africa which is expected to have the fastest growth rate at 2.4% p.a. The Asia and Pacific region will account for nearly half the world's population, and India, with an additional 138 million people by 2027, should overtake China as the most populous country.

Among OECD countries, the population of Japan is expected to decrease by more than 4 million over the next ten years and that of the Russian Federation by 2.1 million. The population of the European Union is expected to remain stable. Australia has the highest projected population growth among OECD countries at 1.1% p.a., followed by Mexico at 1.1% p.a.

Inflation

Inflation rates are projected to increase over the next few years in both advanced and emerging market and developing economies, reflecting the recovery in demand and the increase in commodity prices, including energy prices. Inflation increased in OECD countries in 2017, averaging around 2%, but remained weak in Australia and Canada at around 1% and was close to zero in Japan.

Inflation is projected to increase gradually in the United States, averaging 2.3% p. a. during the next ten years. For the EU15 members as a group, the annual average inflation rate is projected at 1.8% for the next ten years. Inflation is expected to increase slightly in Japan, averaging at 1.6% p.a. Among the major emerging economies, consumer price inflation is projected to remain stable in China at around 2.6% p.a. on average over the projection period, and ease slowly in Brazil at 4.1% p.a., while inflation in the Russian Federation rates should decline to 4.0% p.a. on average.

Exchange rates

Nominal exchange rates for the period 2018-27 are assumed to be mostly driven by the inflation differential in relation to the United States (with minor or no changes in real terms).

The Euro appreciated slightly in nominal terms against the US dollar in 2017 and should appreciate more in 2018, before depreciating again over the next ten years. The currencies of China and Japan are expected to appreciate in nominal terms relative to the US dollar over the next ten years. Conversely, strong depreciation is projected in the currencies of Argentina, Brazil, India, South Africa, Turkey, Paraguay and Nigeria, with lesser depreciations in Korea, Australia, Mexico, the Russian Federation, and Canada.

Energy prices

Historical data for world oil prices to 2016 is based on Brent crude oil prices obtained from the short-term update of the *OECD Economic Outlook* N°102 (November 2017). For 2017, the annual average monthly spot price in 2017 was used, while the estimate for 2018 is based on the average of daily spot prices in December 2017. Oil prices during the projection period follow the path of the World Bank average crude oil price projected by the World Bank Commodities Price forecasts, released in October 2017.

In 2017, crude oil prices started to recover following the extension of the production agreement by the Organization of the Petroleum Exporting Countries (OPEC). Despite strong shale production in the United States, oil prices are projected to continue to increase moderately over the next few years. The baseline projections assume that nominal oil prices increase at an average annual rate of 1.8% over the outlook period, from USD 54.7 per barrel in 2017 to USD 76.1 per barrel by 2027. (Implications of an alternative oil price scenario are explored in Box 1.3).

Policy considerations

Policies play an important role in agricultural, biofuel and fisheries markets, with policy reforms often changing the structure of markets. This *Outlook* assumes that policies will remain as they are throughout the projection period. The decision by the United Kingdom to exit the European Union is not included in the projections, as the terms of that departure have not been determined. In the current *Outlook*, projections for the United Kingdom are retained within the European Union aggregate. In the case of bilateral trade agreements, only ratified or implemented agreements are incorporated. Thus, the North American Free Trade Agreement (NAFTA) remains unchanged throughout the *Outlook* projection period, while the partly implemented but not yet ratified Comprehensive Economic and Trade Agreement (CETA) is incorporated. The Comprehensive and Progressive Agreement for Trans-Pacific Partnership (CPTPP), which was signed in March 2018 and replaced the Trans-Pacific Partnership (TPP) following the withdrawal of the United States, has not been ratified and is not included. The ban by the Russian Federation on imports originating from specific countries was announced as a temporary measure and this *Outlook* assumes that the ban will be revoked at the end of 2018. The specific assumptions on biofuel policies are elaborated in the biofuel chapter.

Reference

Araujo-Enciso, S.R., S. Pieralli, and I. Pérez Domínguez (2017), "Partial Stochastic Analysis with the Aglink-Cosimo Model: A Methodological Overview", *JRC Technical Report*, EUR 28863 EN, doi:10.2760/680976

Chapter 2. The Middle East and North Africa: Prospects and challenges

This chapter reviews the prospects and challenges facing the agricultural sector in the Middle East and North Africa (MENA) region. A dominant concern in the MENA region is its high and growing dependence on international markets for key staple food products, as arable land and water grows scarcer. Policies in the region support grain production and consumption, with the result that 65% of cropland is planted with water-thirsty cereals, in particular wheat which accounts for a large share of calorie intake. The outlook for the MENA region projects slow growth in food consumption, gradual changes in diet to include higher livestock consumption, continued water use at unsustainable rates, and continued and increasing reliance on world markets. An alternative approach to food security would reorient policies towards rural development, poverty reduction, and support for production of higher-value horticulture products. Such a change would contribute to more diversified and healthier diets, but would require building the capacity of farmers to minimise risk while raising higher value crops

Introduction

The Middle East and North Africa (MENA) region[1] consists of a heterogeneous group of countries ranging from the high-income oil-exporting countries in the Gulf, to middle-income and lower middle income countries as well as least developed countries such as Sudan, Yemen and Mauritania (Table 2.1, col. 1). As one of the largest global net food importing regions, it faces considerable uncertainties on both the supply side and the demand side. The former include limitations on, and sustainability of, the production base. Demand side concerns include the impact of ongoing geopolitical conflict, instability in global oil markets which provide the primary source of economic wealth in the region, and rising diet and nutritional issues.

A dominant concern in the MENA region is its high and growing dependence on international markets for key staple food products. This concern has led to a suite of policies which appear strikingly inappropriate given the region's resources. For example, while MENA is one of the most land and water constrained regions of the world, it has the lowest water tariffs in the world and it heavily subsidises water consumption at about 2% of its GDP. As a result, the productivity of water use is only half the world average (World Bank, 2018). Cropping patterns in the region are also difficult to reconcile with the degree of water scarcity. While fruits and vegetables both consume less water and provide higher economic returns per drop, about 60% of harvested land remains in water-thirsty cereals, despite the fact that most countries in the region have a comparative advantage in the export of fruits and vegetables. A key reason for the seeming inconsistency between policy and water scarcity is a vision of food security that aims to reduce dependence on imports, particularly for cereals. At the same time, many countries subsidise the consumption of basic foodstuffs, which in conjunction with rising incomes is contributing to excess consumption of starches and sugars leading to dietary and health concerns such as obesity (FAO, 2017c).

This chapter first considers some of the principal characteristics of agriculture and fish in the MENA region, and reviews performance in terms of resources, production, consumption and trade. The chapter then presents medium-term projections (2018-27) for the agriculture and fish sectors, and then concludes with a discussion of how market balances may evolve, and key risks and uncertainties that may affect this assessment.

The context

Despite their heterogeneity, countries in the MENA region share a number of characteristics, highlighted in Table 2.1. Growth in the region has underperformed, with GDP per capita growing at only 1.6% per year from 2001 to 2016, while middle income countries overall grew by 4.3% p.a. over the same period (col. 2). This is partly due to relatively high population growth in the region which was still over 2% p.a. in the last decade, higher than the global average growth rate of middle income countries during that time of 1.3% p.a. The region also suffers from severe land constraints. Less than 5% of land is arable in two-thirds of the countries of the region, while many countries (Saudi Arabia, Lebanon, Tunisia, Morocco, Yemen, Mauritania and Syria) have huge desert pastures for livestock grazing. The region is the most water-stressed in the world, and two-thirds of countries continue to use groundwater at rates exceeding renewable internal freshwater resources (col. 4).[2] Yet the region has the lowest water prices in the world, spends massive resources on water subsidies (about 2% of GDP) and has total water productivity of only half the world average (World Bank, 2018).

Table 2.1. Contextual indicators for the Middle East and North Africa, 2014

	GDP per capita		Agricultural land	Arable land	Renewable internal freshwater resources	Annual freshwater withdrawals	Exports (2014)	Imports (2013)
	Current USD*	Growth in % per year, 2000-16	% of total land area (2014)		(2014) billion m3		Share of mineral fuels, lubricants and chemical products (%)	Self-sufficiency ratio (%)
	(1)	(2)	(3)		(4)		(5)	(6)
Qatar	86 853	0.6	6	1	0.06	0.44	87	3
United Arab Emirates	44 450	-2.1	5	0	0.15	4.00	38	
Kuwait	42 996	0.1	9	1	0.0	0.9	94	
Bahrain	24 983	-0.1	11	2	0.0040	0.3574	48	
Saudi Arabia	24 575	1.2	81	2	2	24	90	33
Oman	20 458	-0.2	5	0	1.40	1.32	79	5
Lebanon	8 537	0.4	64	13	4.8	1.3	13	41
Iraq	6 703	2.7	21	12	35	66	95	54
Libya	5 603	-2.4	9	1	0.7	5.8	77	
Iran, Islamic Rep.	5 541	2.5	28	9	129	93	77	85
Algeria	5 466	2.0	17	3	11	8	98	64
Tunisia	4 270	2.3	65	19	4	3	14	75
Jordan	4 067	1.1	12	3	0.7	0.9	32	38
Egypt, Arab Rep.	3 328	2.2	4	3	2	78	31	72
Morocco	3 155	3.0	69	18	29	10	16	80
Palestinian Authority	2 961	0.6	50	11	0.81	0.42	6	16
Sudan	2 177	4.2	29	8	4	27	64	85
Syrian Arab Republic	2 058	2.1	76	25	7	17	24	
Yemen, Rep.	1 647	-2.4	45	2	2	4	41	50
Mauritania	1 327	1.4	39	0.4	0.4	1.4	4	

Note: All GDP per capita estimates are for 2014, except for Libya (2011) and Syria (2007), for which conflict affected availability of reliable data. GDP per capita growth for Syria is 2000-2007, and for Libya, 2000-11. Arable land includes land under temporary crops, temporary meadows, kitchen gardens and land temporarily fallow. Agricultural land includes arable land, as well as land under permanent crops, and under permanent pastures. The self-sufficiency ratio for Table 2.1 is in value terms: (value of gross agricultural production in current US dollars)*100/(value of gross agricultural production in current US dollars + value of imports in current US dollars – value of exports in current US dollars).
Source: World Bank (2018); UNCTAD (2018); FAO (2018a, 2018b).

The scope of merchandise exports from the region remains limited, with over two-thirds of exports consisting of mineral fuels, lubricants and chemical products (col. 5). This narrow range of products makes exports from the MENA region nearly ten times more concentrated than in the rest of the world. Whereas the concentration index of exports in the world was 0.06 in 2014, the index was 0.44 in the MENA region (UNCTAD, 2018).[3] However, there is a great diversity in the reliance on petroleum exports in the region. Such countries as Iraq, Algeria, Saudi Arabia, Qatar and Kuwait export little else but mineral products, lubricants and chemicals, while Mauritania, the Palestinian Authority, Lebanon, and Morocco export very few of such products.

Finally, though the region has dramatically increased its participation in global agricultural markets as a share of GDP in the past 50 years, this surge was predominantly due to rising imports. In 2013, domestic agricultural production accounted for 65% of the value of agricultural products consumed domestically, though this share varied from 3% in Qatar to 85% in Sudan and Iran (col. 6). The remaining agricultural products were supplied from imports.

Agricultural use of natural resources in the MENA region

The MENA region is a difficult environment for agriculture. Land and water are scarce, and both rain-fed and irrigated land in use suffer from ongoing degradation caused by wind and water erosion and unsustainable farming practices. In most countries, farms are quite small and hence subject to the challenges experienced by smallholders everywhere. Furthermore, the region is predicted to become hotter and drier in the future due to climate change.

Only a small share of land in the region is arable

Of the total land area of the MENA region, only one-third is agricultural land (cropland and pastures), while only 5% is arable (cropland) (Table 2.1). The rest of the land is either urban or dry desert. Due to the dry climate, about 40% of cropped area in the region requires irrigation (FAO, 2018a, 2018b). Figure 2.1 shows that only 4% of land in the region has soils judged of high or good suitability for rain-fed cereal cultivation and 55% is unsuitable.

Figure 2.1. North Africa and West Asia crop suitability index (class) for low-input rain-fed cereals, 1961-1990

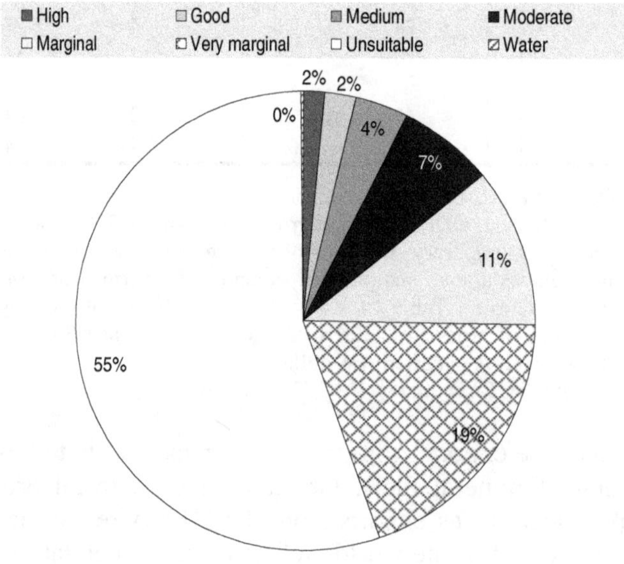

Source: FAO (2018c).

StatLink http://dx.doi.org/10.1787/888933742454

In addition to the dearth of suitable land for cultivation, soils currently used for farming are severely degraded to the point where their productivity is estimated to have been reduced by up to 30 to 35% of potential productivity (Box 2.1). Soil degradation in rain-fed systems is caused by wind and water erosion, while in irrigated systems the farming practices themselves are responsible for soil salinity and sodicity.[4] Three-quarters of the region's 30 million ha of rain-fed cropland are estimated to be degraded. Recent studies have estimated the economic cost of land degradation in the region at USD 9 billion each year (between 2% and 7% of individual countries' GDP). Losses from salinity alone across the region are estimated at USD 1 billion annually, or USD 1 600 to USD 2 750 per ha of affected lands (ESCWA and FAO, 2018).

Box 2.1. Initiatives to address land quality issues in the MENA region

Zero tillage. Ploughing up soil can have many deleterious effects, such as loss of moisture and organic matter, which increase the vulnerability to wind and water erosion. Farming with zero or minimum tillage can avoid these problems by eliminating ploughing, leaving the soil undisturbed. Roots left from the previous crop stabilise the soil, hence protecting against erosion, while the organic matter above ground adds to the fertility and water holding capacity of the soil. Seed drills are used to insert seeds and fertiliser directly into the soil without ploughing. However, seed drills are expensive, and most smallholder farms are not able to afford the cost of about USD 30 000. A recent project by ICARDA and the Australian government has addressed this problem. Working with local farmers and craftsmen, the project has produced and distributed almost 200 affordable seed drills which are now being used across the Syrian Arab Republic, Iraq, Lebanon, Jordan, Algeria, Tunisia and Morocco.

Soil maps. Soil data is important for farmers and policymakers. However, soil maps are often outdated, of low resolution and not easily understandable. The Amman-based Institute of Digital Soil Mapping is serving as a regional hub for a global consortium of scientists and researchers. The consortium is developing GlobalSoilMap.net, which can combine data from several sources and present it in a user-friendly format for a broad range of audiences. The data can include soil pH, water storage electrical conductivity and carbon content data derived from remote sensing, near- and mid-infrared spectroscopy and field sampling. The initiative can also make use of the Global Soil Partnership system of the International Network for Soil Information Institutes. In addition, the European Union, the African Union and FAO have recently published a Soil Atlas of Africa (Jones et al., 2013).

Sources: www.icarda.org/conservation-agriculture/zero-tillage-seeders, cited in ESCWA and FAO (2018).

Land productivity is low compared to other regions

An overall indicator of the productivity of land use is the value of gross agricultural production per ha of agricultural land, which is lower in MENA than in most areas of the world (Table 2.2). [5] Of the major regions, only Sub-Saharan Africa has a worse performance. The low value of production per hectare reflects the high share of arable land devoted to low-yield temperate crops, as well as the low productivity of desert pastures. Not all countries perform so poorly. Egypt, with rich soils, irrigated cereals production and virtually no pastures, produces over USD 6 000 worth of products on each hectare of agricultural land, while Bahrain, which produces only horticultural crops and livestock, produces over USD 4 000 worth of product. Jordan, Lebanon, the Palestinian Authority, the UAE and Kuwait also produce over USD 1 000 worth of product per hectare, with very little area devoted to cereals.[6]

Table 2.2 also allows a comparison of the growth in land productivity in the MENA region versus other developing regions. While progress was good in the 1970s, the relative performance of MENA has been less impressive in more recent decades. Since the 1980s, decade to decade growth in the MENA region has ranked at the bottom of the four developing regions in Table 2.2, indicating a relative deterioration of its performance compared to other developing regions.

Table 2.2. Value of gross production per hectare of agricultural land (constant 2004-2006 prices in thousands of international dollars per year)

	1961-70	1971-80	1981-90	1991-00	2001-14
World	189	234	286	334	449
Western Europe	1 284	1 541	1 810	1 878	1 962
North America	261	326	375	449	540
East Asia	209	269	364	518	829
Latin America and Caribbean	138	169	213	258	373
Sub-Saharan Africa	55	67	79	104	146
MENA	85	111	142	162	226

Source: FAO (2018b).

For horticultural crops (such as oranges and tomatoes) the MENA region has yields similar to the world average. However, average yields of temperate crops such as wheat and oilseeds are far below world levels (Table 2.3). This low average hides differences across countries however, as yields differ depending on irrigation and the application of fertiliser and other inputs. Egypt, Kuwait, Saudi Arabia, the UAE, Oman and Lebanon all achieved wheat yields over 3 tonnes per ha in 2010-16 (Figure 2.2). Each of these countries has irrigated wheat production and applied between 100 kg and 600 kg of fertiliser (in nutrient weight terms) per ha of arable land per year in the period 2010-15 (FAO, 2018b).

Production of horticultural crops and cereals has increased over the period 1971-2016 through both area expansion and higher yields. This is not the case for oilseeds, where production declined over time. For oranges, tomatoes and wheat, yields in MENA have grown at a slightly higher rate than the world average. Moreover, the growth in area has been stronger for horticultural crops than for temperate crops such as wheat and oilseeds (Figure 2.5).

Table 2.3. Average yield of oranges, tomatoes, wheat and oilseeds, by region, 2010-16 (tonnes per ha)

	Oranges	Tomatoes	Wheat	Oilseeds
World	17.9	35.2	3.2	3.2
Western Europe	5.8	269.5	7.2	3.2
North America	28.3	91.1	3.1	2.0
East Asia	15.3	52.1	5.0	2.8
Latin America and Caribbean	19.3	38.7	3.1	4.5
Sub-Saharan Africa	17.6	7.8	2.5	1.8
MENA	17.9	37.8	2.2	0.9

Source: FAO (2018b).

Figure 2.2. Average wheat yield in the MENA region, by country, 2010-16

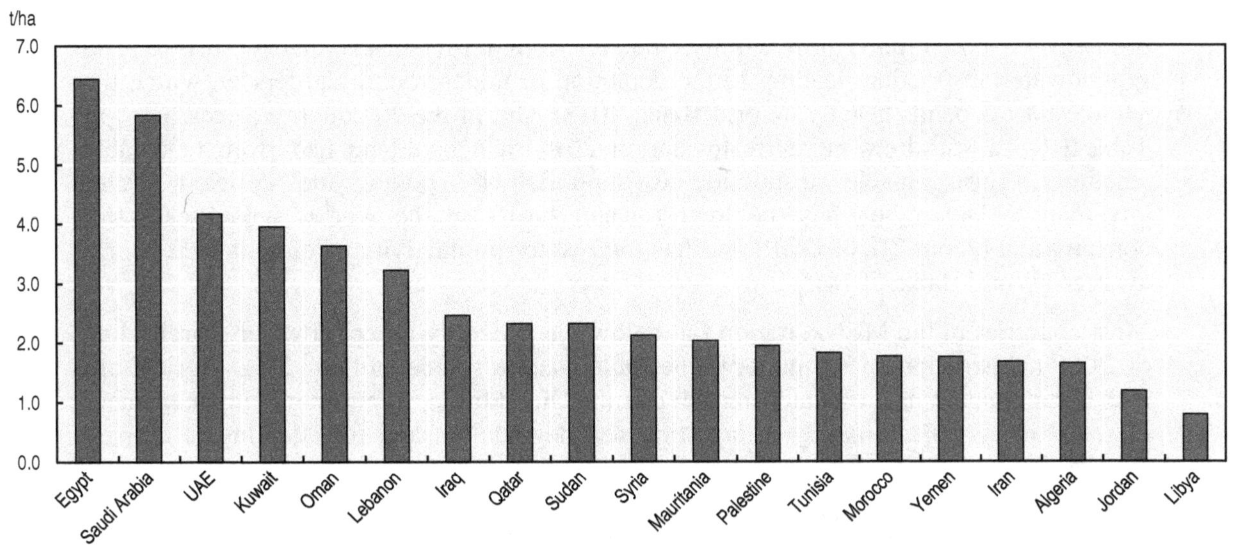

Note: Saudi Arabia was largely out of wheat production by 2015.
Source: FAO (2018b).

StatLink http://dx.doi.org/10.1787/888933742473

Table 2.4. World and MENA: Average annual growth in production, yield and area for oranges, tomatoes, wheat and oilseeds, 1971-2016 (%)

	Oranges	Tomatoes	Wheat	Oilseeds
World				
Production	2.3	3.5	1.7	4.4
Yield	0.4	1.4	1.7	2.2
Area harvested	1.9	2.1	0.1	2.2
MENA				
Production	3.1	4.2	2.4	-1.0
Yield	0.6	2.5	2.2	-1.2
Area harvested	2.5	1.6	0.2	0.2

Source: FAO (2018b).

As discussed in more detail below, farms are quite small in most countries in the region, and these small farms tend not to specialise. They have a comparative advantage in labour-intensive horticultural crops, since they have plentiful household labour, but are limited in their ability to adopt new technology and access investment. Moreover, smallholders are averse to specialising in horticulture because of the higher risks involved. Horticultural crops have potentially high payoffs, but also have higher input costs; in a bad year, a farm can lose its entire investment on seeds, fertiliser and pesticides. By contrast, cereals are more robust, low-input, low-yield crops. Smallholders thus often cultivate both horticultural crops and cereals as a diversification strategy to lower their risk, ensure a minimum income and provide for direct consumption. In combination with poor natural growing conditions, the low degree of specialisation contributes to lower yields in both horticultural and cereal crops. The low productivity of smallholder farms in the MENA region is consistent with this analysis.

Agricultural water policies and use are increasingly unsustainable

It is difficult to overestimate the importance of the water issue in the MENA region. Along with conflict, it is the most profound man-made threat to the region's future. The problem is not limited to scarcity, but of long-term unsustainable surface and groundwater abstraction, leading to the depletion of underground aquifers on which the Middle East depends heavily (World Bank, 2018). Out of the 20 countries/areas listed in Table 2.1, 13 withdrew more freshwater in 2014 than could be had from renewable resources. Unsustainable abstraction is supported by policy and deficient water governance. The region has the lowest water tariffs in the world, subsidises water consumption (about 2% of GDP) and has total water productivity of only half the world average (World Bank, 2018).

Most countries in the MENA region fall below the generally accepted water scarcity line of 1 000 m3 per capita per annum of renewable water resources (Figure 2.3).[7] Agriculture is the predominant user of water in each country. In addition, improving the management of water in agriculture is key to arresting soil degradation and for adapting to climate change.

Figure 2.3. Annual renewable water resources per capita, 2014

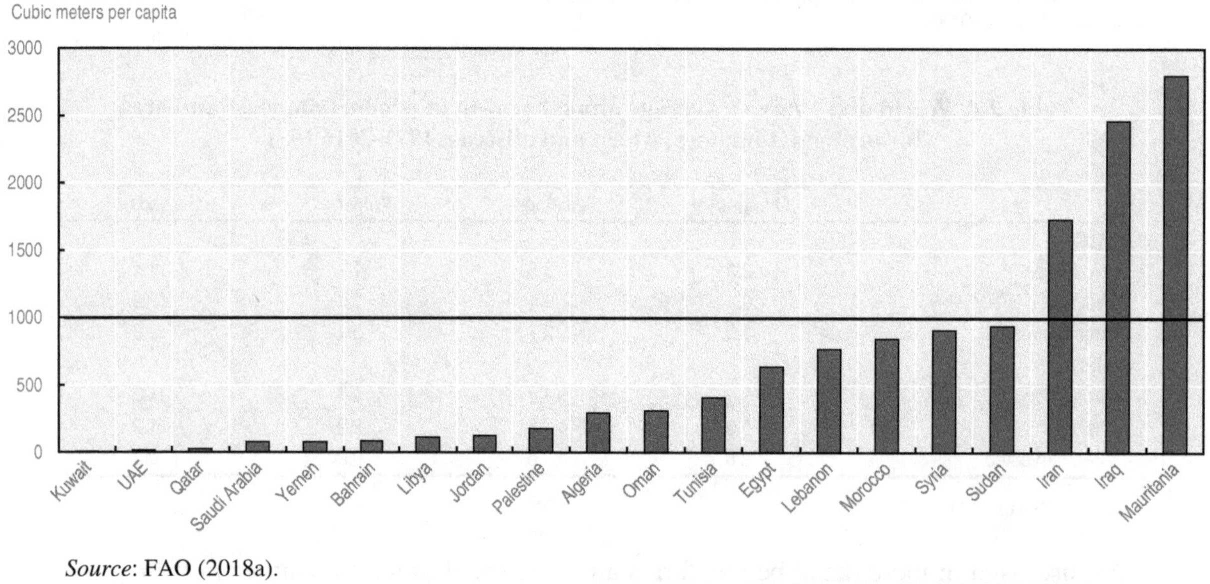

Source: FAO (2018a).

StatLink http://dx.doi.org/10.1787/888933742492

Water productivity is one of the main concerns in MENA agriculture

The productivity of water used in agricultural production may be measured in two main ways.[8]

- *Physical water productivity* is the *volume* of agricultural production per unit of water consumed in the production of that output. Table 2.5 (col. 1) illustrates that in the MENA region the physical water productivity is highest for vegetables and fruits, followed by cereals, groundnuts and livestock products. There is a wide range of physical water productivities for each product, because of differences in soil fertility, plant disease, pests, and the timing of watering and planting, which all influence water productivity. The more a farmer can control these factors (e.g. through irrigation, proper agronomic practices, fertilisation and control of plant disease and pests), the higher the physical water productivity that can be attained.

- *Economic water productivity* may be defined as the *value* of production per unit of water used. In the MENA countries, the highest value per cubic meter of water used is obtained for vegetables and fruits, followed by olives, dates, lentils, cereals, and beef (Table 2.5, col. 3).

Table 2.5. Average water productivity for selected agricultural products in the MENA region

	Physical water productivity, midrange value (kilograms per M3)*	Average producer price in MENA, 2010-16 (USD per kg)**	Average economic water productivity (USD per M3 of water in producing agricultural commodity)
	(1)	(2)	(1)*(2)=(3)
Tomato	12.5	0.40	4.98
Onion	6.5	0.42	2.76
Apples	3.0	0.88	2.64
Potato	5.0	0.45	2.23
Olives	2.0	0.90	1.80
Lentils	0.7	1.17	0.82
Dates	0.6	1.33	0.80
Fava beans	0.6	0.98	0.54
Maize	1.2	0.45	0.51
Rice	0.9	0.59	0.51
Bovine meat	0.1	7.48	0.49
Wheat	0.7	0.51	0.33
Groundnut	0.3	1.33	0.33

Note: *Calculated as mean of minimum and maximum from Molden, et al., 2010. **MENA country average, 2010-16, from FAO (2018b).
Source: Molden et al. (2010); FAO (2018b).

Water is not the only input in agricultural production, and other factors influence the decision of which crops or livestock to produce. Decisions on product selection also depend on the type of land available (e.g. pasture *vs.* cropland), the location of the farm (e.g. in rain-fed or irrigated areas), and farmers' attitudes towards risk. However, if other costs are similar, a farmer in the MENA region would obtain the highest payoff per drop of water by producing fruits and vegetables.

Impact of climate change on production conditions varies within the region

Climate change in the MENA region only adds to the hazards of farming in an already exceedingly dry area of the world. The MENA countries are prone to frequent droughts and face future water shortages due to unsustainable withdrawal of groundwater. In addition, mean temperatures over the past century have risen by 0.5°C, and precipitation over the past several decades has decreased by up to 10% in some parts of North Africa and Sudan. Climate change projections expect the entire region to become hotter and drier in the future, with a reduction of precipitation particularly evident in the western part of the region (Bucchignani et al., 2018). Higher temperatures and reduced precipitation will accelerate the loss of surface water, and droughts will become more frequent. The already low average yields of rain-fed crops will decline and become more variable. By the end of the century, total agricultural production in the region could decrease by up to 21% from a 2000 base.[9]

While all farming systems will become increasingly arid and water scarce, rain-fed systems are most at risk.[10] However, some areas may benefit from warmer temperatures which extend growing seasons and increase the productivity of winter crops. In Yemen, for example, where there are summer rains, an increase in average temperatures of 2ºC could be expected to extend the growing season by about six weeks (Verner and Breisinger, 2013). Furthermore, some areas are expected to receive more rainfall, which may raise yields, though they may also increase the frequency of floods. These trends have already been observed in Oman, Saudi Arabia and Yemen.

The common denominator of climate change will be a general increase in temperature in this region with varying rainfall effects across countries. However, the effects of climate change on agriculture are expected to vary by farming system (Table 2.6). In some cases, farmers can respond to changes through adaptation. In other regions, agriculture may become untenable, and rural inhabitants will need to transition to off-farm employment or relocate.

Table 2.6. Climate change impact on farming systems in the MENA region

Farming system	Exposure: Expected climate change-related events	Sensitivity: Likely impact on farming systems
Irrigated	Increased temperatures Reduced supply of surface irrigation water Dwindling of groundwater recharge	More water stress Increased demand for irrigation and water transfer Reduced yields when temperatures are too high Salinisation due to reduced leaching Reduction in cropping intensity
Highland mixed	Increase in aridity Greater risk of drought Possible lengthening of the growing period Reduced supply of irrigation water	Reduction in yields Reduction in cropping intensity Increased demand for irrigation
Rain-fed mixed	Increase in aridity Greater risk of drought Reduced supply of irrigation water	Reduction in yields Reduction in cropping intensity Increased demand for irrigation
Dryland mixed	Increase in aridity Greater risk of drought Reduced supply of irrigation water	A system very vulnerable to declining rainfall Some lands may revert to rangeland Increased demand for irrigation
Pastoral	Increase in aridity Greater risk of drought Reduced water for livestock and fodder	A very vulnerable system, where desertification may reduce carrying capacity significantly Non-farm activities, exit from farming, migration

Structure and performance of agriculture, fisheries and aquaculture in the Middle East and North Africa

Uneven farm size distribution across the region

The Middle East and North Africa has one of the most uneven farm size distributions in the world. In some of the countries in the region — Egypt, Yemen, Jordan, Lebanon and Iran — the majority of farms are smaller than one hectare (Figure 2.4). At the other end of the size spectrum are a relatively small number of large farms owned by a small number of landowners or the state (Rae, n.d.).

The inequality of landholding is illustrated in Figure 2.5 using Lorenz curves, which plot the cumulative share of farms against the cumulative share of agricultural land. The diagonal line illustrates a theoretical case in which each holding is of an equal size such that, for instance, 50% farms occupy 50% of total agricultural area. The more bowed-out the actual Lorenz curve, the more unequally holdings are distributed. For instance, 80% of farms occupy only 20% of total agricultural area in the Middle East and North African region, indicating that the overwhelming majority of farms are quite small. On the other hand, another 10% of farms holds 60% of agricultural area, implying that a small number of large farms cultivate over half of agricultural land area. Only in Latin America, the distribution of land is even more unequal: less than 10% of farms hold 80% of agricultural land area.

Two policies can be observed in the MENA region that support the concentration of farmland through supporting the development of large-scale farm enterprises. First, the predominant policy in the region for the development of rural areas is the sectoral modernisation of agriculture, which includes the promotion of large intensively cultivating corporate or private farms. Public support to agriculture and access to credit de facto favours large farms, often for sound business reasons. Due to their size, small farms are often not eligible to benefit from public support or bank loans. Sectoral "modernisation" policies have largely excluded smallholders from public support, which have left them small, technologically backward and poor. Alternative policies of rural development focused on supporting small farms through technical and business training, and small and medium rural enterprise and community development is often absent or poorly funded.

A second policy that supports the concentration of holdings in large farms is state facilitation for the large-scale acquisition of land by both domestic and foreign investors. This policy has been pursued most intensively in Sudan and Egypt, though land has also been made available in Mauritania or Morocco. In the MENA region most land acquisitions have been pursued by corporations with the support of governments and banks from water-scarce, wealthy GCC (Gulf Cooperation Council) countries with the largest dependence on food imports. Foreign land acquisition in the region developed during the 2007-2014 period of high commodity prices, and is aimed at limiting exposure to world commodity markets and ensuring access to food and feed supply in the GCC countries. Case studies from Sudan indicate that the terms of large-scale purchase or leasing contracts often lack transparency, and are reached with little or no consultation with local communities. Large tracts of communal land in Sudan were sold or leased to local or foreign investors, with little attention to the social cost and environmental impacts from turning communal pasture land into foreign owned cropland (Elhadary and Abdelatti, 2016).

Figure 2.4. Farm size distribution in selected MENA countries, 1996-2003

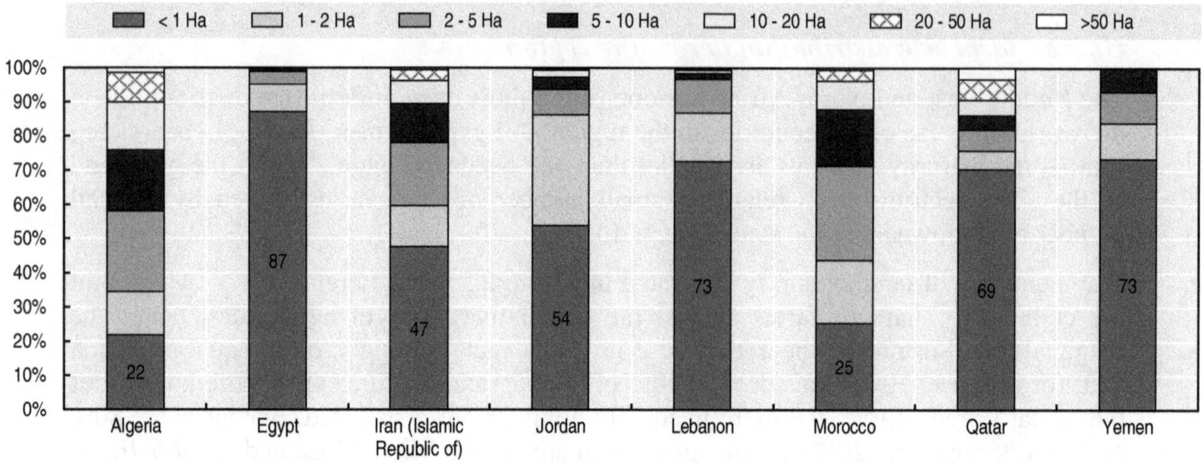

Note: Figures in the <1 ha portion of the bars show the share of holdings of less than 1 ha. Estimates refer to the size distribution of holdings in Algeria (2001), Egypt (1999-2000), Iran (Islamic Rep. of) (2003), Jordan (1997), Lebanon (1998), Morocco (1996), Qatar (2000-2001) and Yemen (2002). The figures in the bars indicate the share of holdings under 1 ha.

Source: Lowder et al. (2014).

StatLink http://dx.doi.org/10.1787/888933742511

Figure 2.5. Concentration of agricultural land in farm holdings: MENA in comparative perspective

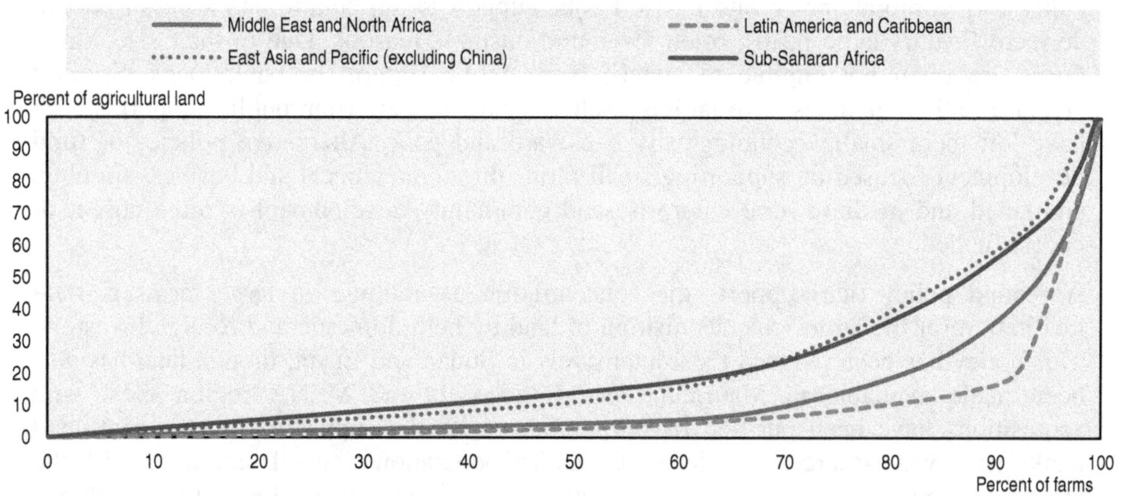

Source: Lowder et al. (2014).

StatLink http://dx.doi.org/10.1787/888933742530

Agricultural production dominated by cereals

The scarcity of water, the shortage of suitable land and the constraints of smallholder farming all impact on production in the MENA region by limiting yields. Low-yield agriculture in the region is characterized by low diversity such that harvested area is dominated by cereals (Figure 2.6).[11] Cereals occupied about 60% of the harvested land area in the region, but contributed only 15% of the value of gross agricultural production in 2014. Cereal production has been encouraged by policies to lower import dependence.

Figure 2.6. MENA Harvested area, share by commodity type, 1961-2016

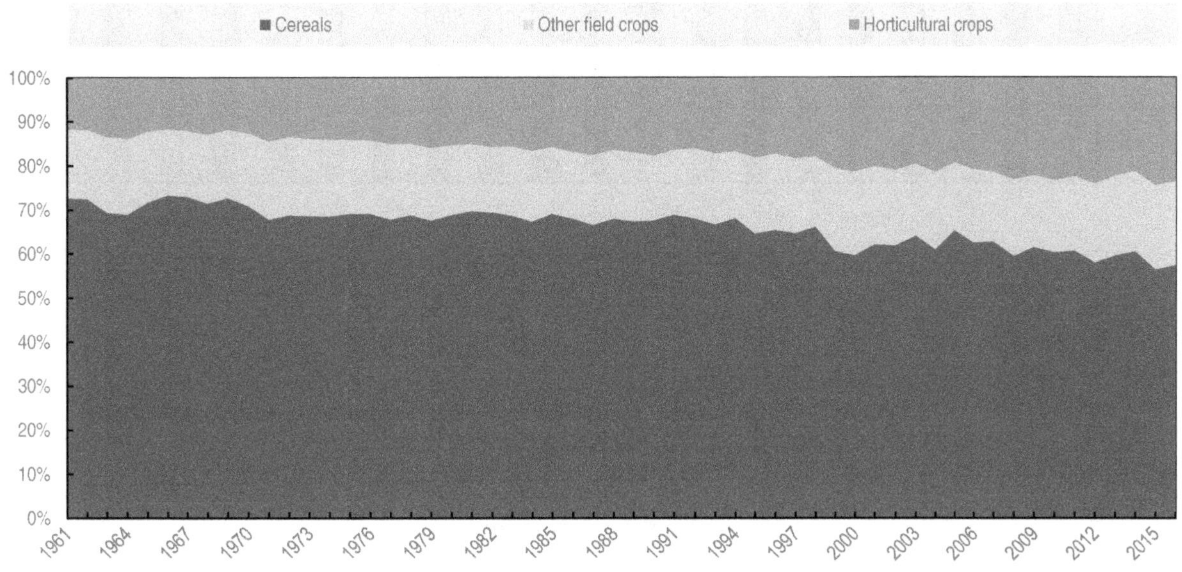

Note: Horticulture includes citrus, fruits, berries, vegetables, melons, tree nuts, herbs, tea, coffee, spices, stimulants, beverage crops and olives. Other field crops include fibres, beans, peas, sugar crops, roots and tubers, pulses, and oilseeds.
Source: FAO (2018b).

StatLink http://dx.doi.org/10.1787/888933742549

Although cereals occupy about 60% of total harvested area, this share varies widely by country (Figure 2.7). The poorer countries, such as Sudan, Yemen, Iraq and Mauritania, devoted most of their land to cereals. However, other countries, including those in the GCC, Lebanon, Tunisia, Libya, the Palestinian Authority, and Jordan, devoted over 50% of harvested area to horticultural crops, and cereal production is low.[12]

While land area in the region is dominated by cereals, most of the value of production in the region comes from horticultural crops and livestock (Figure 2.8). Generally, about 40% of the value of agricultural production now comes from horticulture.

Finally, MENA agriculture is dominated by two regional giants (Iran and Egypt), which together produce half of the total value of agricultural production (Figure 2.9). The next three producers by size are Sudan, Morocco and Algeria, which together produce 27% of agricultural production. The remaining 15 countries produce 23% of the total value of agricultural production in the MENA region.

Figure 2.7. MENA Harvested area share, by country and crop type, 2016 (percent)

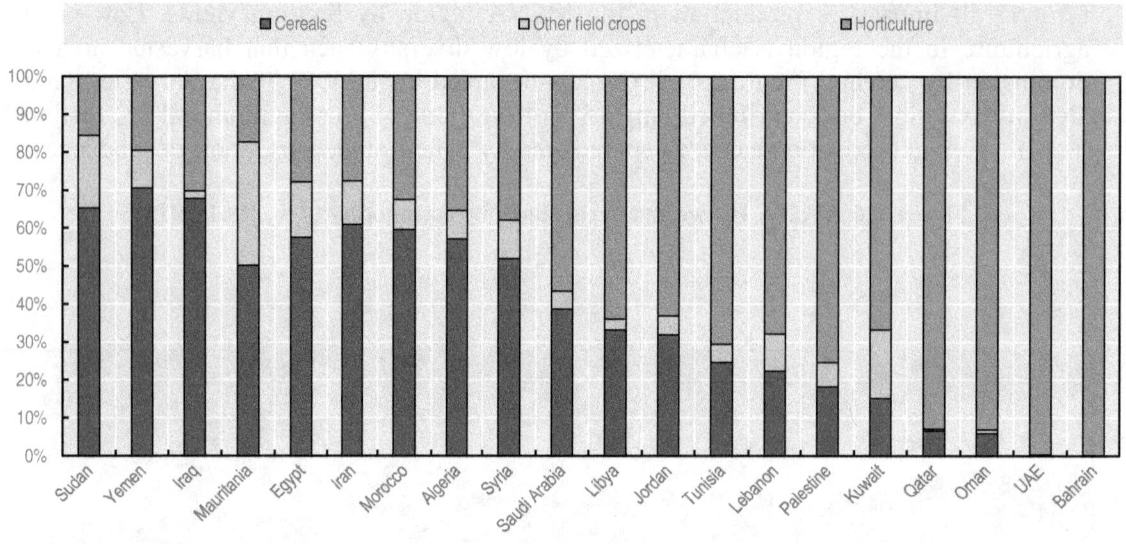

Source: FAO (2018b).

StatLink http://dx.doi.org/10.1787/888933742568

Figure 2.8. MENA value of agricultural production, share by commodity type, 1961-2014, percent

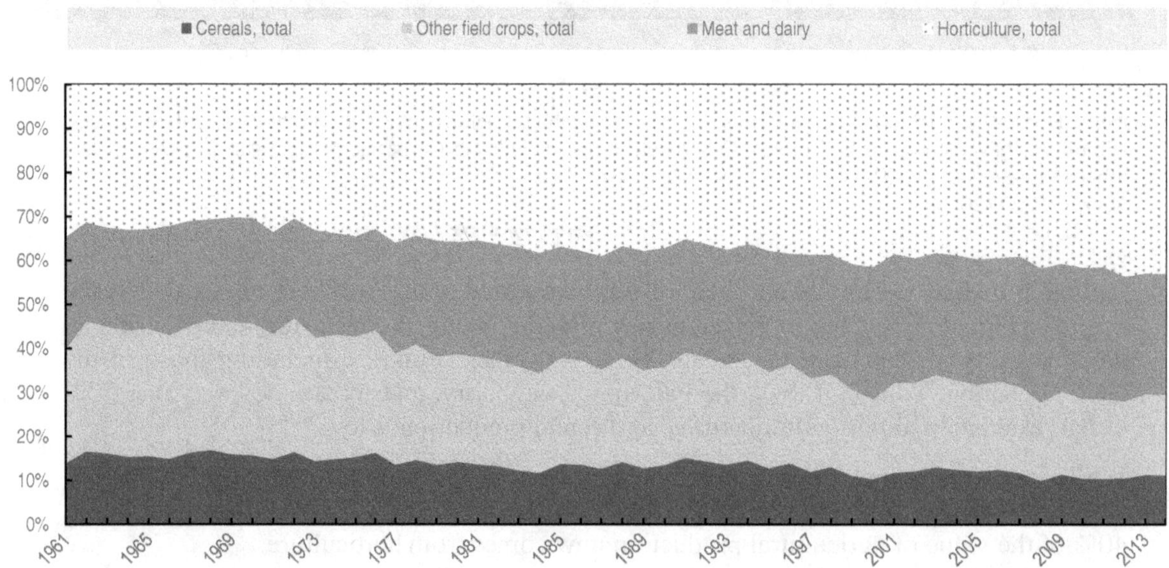

Note: Horticulture includes citrus, fruits, berries, vegetables, melons, tree nuts, herbs, tea, coffee, spices, stimulants, beverage crops and olives. Other field crops include fibres, roots and tubers, beans, peas, pulses, sugar crops and oilseeds.
Source: FAO (2018b).

StatLink http://dx.doi.org/10.1787/888933742587

Figure 2.9. The value of agricultural production in the MENA region, by country and commodity type, 2014

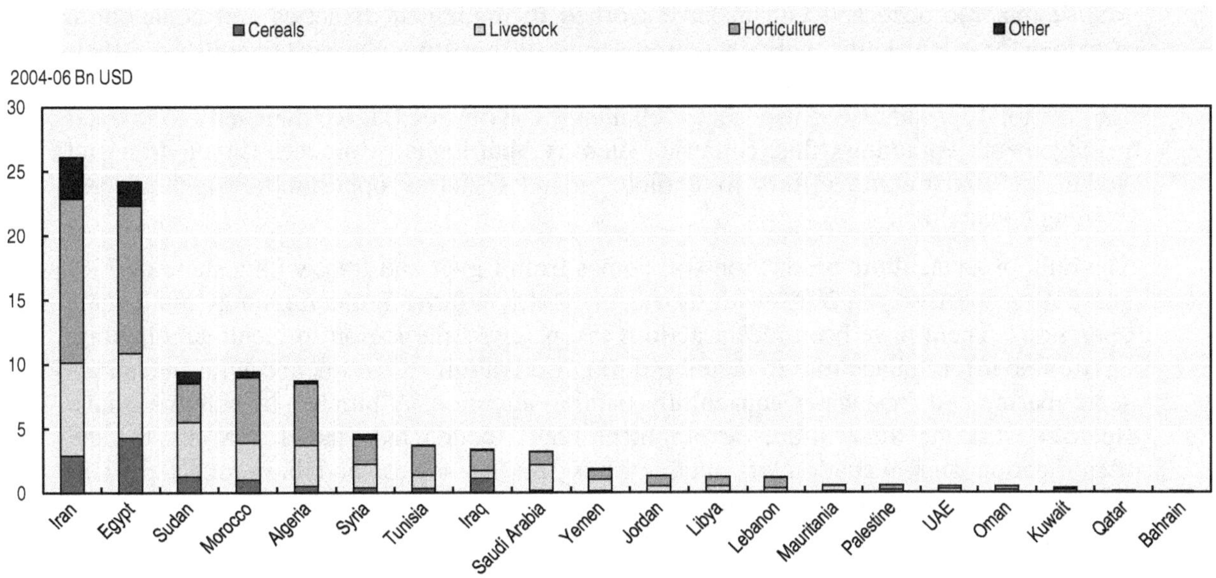

Note: Data for Syria in 2014 may not be reliable.
Source: FAO (2018b).

StatLink http://dx.doi.org/10.1787/888933742606

Fisheries and aquaculture in the MENA region

The MENA region includes diverse marine and freshwater ecosystems. Although the region is generally arid, it also encompasses major transboundary waterways such as the Euphrates, the Tigris, the Nile, and other river systems. However, overall freshwater resources remain scarce, particularly in areas away from river systems. Capture fisheries and aquaculture are important in the MENA region as providers of livelihoods and as sources of nutritious food. During the last two decades, total capture fisheries and aquaculture production increased significantly from 2.2 Mt in 1996 to 5.9 Mt in 2016. Most of the increase originated from capture fisheries (from 2.0 Mt to 4.0 Mt), but aquaculture registered strong growth as well (from 0.1 Mt to 1.9 Mt), with its share in total fish production increasing from 6% to 32% during the 1996-2016 period. Despite this increase in production, the region is dependent on imports of fish and fish products to satisfy domestic consumption.

The aquaculture and fisheries sector in the MENA region faces many challenges with marked differences among and within countries. Marine capture fisheries in the MENA coastal countries range from the large annual production of countries with long coastlines and large fleets that access highly productive upwelling systems, to the countries with smaller production, and smaller fleets. Coastal areas across the region are important for small-scale fisheries that support the livelihoods of hundreds of thousands of people and overall fisheries are overwhelmingly small scale. Biomass assessments, undertaken on only a limited number of the main stocks fished throughout the region, found that most are under pressure. Regional fisheries management organisations (RFMOs), such as the Indian Ocean Tuna Commission (IOTC) and the International Commission for the Conservation of Atlantic Tunas (ICCAT), are implementing adaptive management measures to maintain stocks within safe biological levels and the Regional Commission for Fisheries (RECOFI) recently adopted binding recommendations for minimum fisheries and aquaculture data reporting. In addition, many countries in the region, such as Mauritania, Morocco and Oman, have worked to implement fisheries and aquaculture strategies and legislation with increased focus on ensuring the sustainability of their resources. Inland fisheries production, which amounted to 0.4 Mt in 2016, representing 7% of total production, also faces challenges with regard to their environmental management. To address this, countries such as Mauritania, Morocco, Egypt, Iran and Sudan are undertaking efforts to explore inland fisheries opportunities and address existing constraints.

The bulk of aquaculture production still comes from Egypt and Iran, with a share of 73% and 21% respectively in 2016, with the majority of fish farms in the region as small-scale operations. There have been recent actions taken across the region to create an enabling environment for aquaculture to develop through private investments and with industrial-scale marine and freshwater aquaculture gaining attention. A number of countries have finalised strategic aquaculture development plans, conducted spatial analysis for the identification and allocation of suitable sites for the sector, and have enacted clear regulations to assist with the establishment of commercial facilities. The aquaculture sector faces several constraints including limited access to appropriate locations and to sustainable production technologies, inappropriate freshwater fish hatchery installations and management, inadequate seed production in terms of quantity and/or quality and poor handling and transportation. Animal health control systems for aquaculture are also scarce and access to credit, loans and insurance for aquaculture business is almost non-existent in most countries of the region. Furthermore, the expansion of the aquaculture industry in the region has increased environmental concerns and public awareness about food security issues and environmental conservation. In addition, fisheries in the MENA region are particularly vulnerable to the impacts of climate change and variability as well as those induced by human activities. In this respect, the aquaculture sector can be particularly vulnerable as there is a lack of farmer's adaptability to climate change and resilience to natural disasters and socioeconomic risks.

Growing import dependence for basic foods

Low yields and a narrow scope for increases in arable area in the MENA region set limits to crop production for temperate crops, such as wheat and oilseeds. Coupled with income growth and a particularly strong population growth of 2.5% over the period 1971-2016, demand growth has far outstripped production growth for these crops for which the MENA region is ill-suited (Table 2.7). The growing gap between consumption and domestic production (Figure 2.9) has been covered by imports. Growth in horticultural crop production has kept pace with demand, such that the region is self-sufficient in fruits and vegetables (Figure 2.10).

Table 2.7 details that the region is far from self-sufficient in cereals, vegetable oils, oilseeds and sugar and sweeteners, but is self-sufficient or nearly so for fruits and vegetables and meats (including animal fats and offal).

Figure 2.10. Domestic production and use of selected commodities in the MENA region, 1961-2013

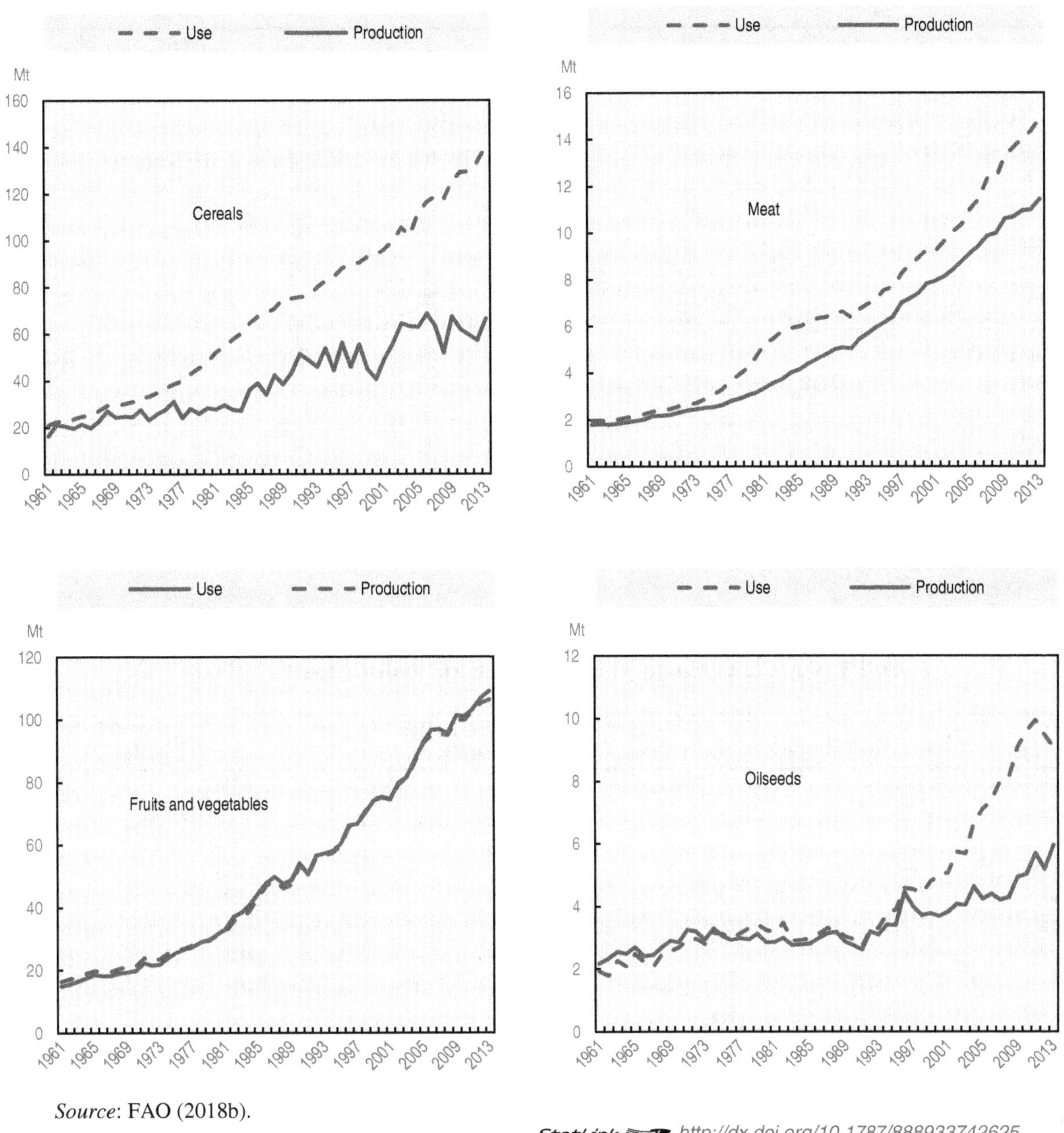

Source: FAO (2018b).

StatLink http://dx.doi.org/10.1787/888933742625

Table 2.7. Food self-sufficiency ratios (SSR) in MENA countries, average, 2011-13 (%)

SSR	Cereals[1]	Meats[2]	Fruits, vegetables	Milk[3]	Vegetable Oils	Oil crops	Sugar, Sweeteners
Algeria	30	91	93	51	11	88	0
Egypt	58	83	107	89	26	35	73
Iran (Islamic Republic of)	61	95	104	106	15	58	58
Iraq	50	34	86	45	2	80	0
Jordan	4	72	139	51	17	80	0
Kuwait	2	34	36	14	1	0	0
Lebanon	14	77	111	49	20	67	0
Mauritania	27	89	18	65	0	95	0
Morocco	59	100	116	95	29	98	28
Oman	7	32	52	32	4	0	0
Saudi Arabia	8	45	73	76	18	1	0
Sudan (2012-13)	82	100	98	96	89	112	72
Tunisia	42	98	110	90	91	65	1
United Arab Emirates	2	26	21	14	82	0	0
Yemen	17	79	90	35	5	63	1
MENA Total	46	79	99	82	25	64	37

Note: The self-sufficiency ratio is defined as food production/(production+imports-exports).
1. Excluding beer.
2. Includes meat and offals.
3. Excluding butter.
Source: FAO (2018b).

Table 2.8. Share of agricultural imports in merchandise exports, 2011-13 (%)

	Agricultural imports as a percentage of merchandise exports (%)	Stability
Total MENA	8	Stable
Palestinian Authority	74	Volatile, 1990-2002
Syria	58	Volatile since 2007
Lebanon	58	Stable
Egypt	49	Stable
Jordan	44	Stable
Yemen	39	Stable
Sudan	34	Stable
Morocco	25	Stable
Mauritania	17	Stable
Tunisia	15	Stable
Algeria	15	Stable
Iran	11	Stable
Libya	9	Stable
Iraq	9	Volatile, 1990-99
Bahrain	8	Stable
Saudi Arabia	6	Stable
Oman	5	Stable
UAE	4	Stable
Kuwait	3	Stable
Qatar	2	Stable

Source: FAO (2018b).

The share of total food imports in total merchandise exports can be used as an indicator to assess the capacity of a country to sustain food imports (Table 2.8). Globally, this share is about 5%. The MENA average has been about 8% in recent years (2011-13), and has shown a downward trend from earlier years. For the countries whose share of total merchandise export earnings spent on food imports is high and volatile, the stability of international food prices is a major concern. Even if export earnings can be maintained, these countries face significant risks associated with spikes in world food prices. The implications of this vulnerability were realised during the 2007-2008 global food crisis, when prices spiked dramatically. The importing countries of the world, including those in the MENA region, were faced with high prices impacting household and government budgets. While world food commodity markets have since returned to more normal conditions, the experience of the crisis brought increased attention to the vulnerabilities of importing countries – and particularly for countries such as the Palestinian Authority and Syria for which food imports constituted a large and volatile share of total export earnings in 2011-13.

The pattern of trade in cereal, oilseed and meat products is consistent with findings based on the Balassa Export Revealed Comparative Advantage Index (XRCA) applied to agricultural products. Table 2.9 shows the comparative strength of exports of six MENA countries in 2011-2013. Though each country is different, most countries have an advantage in the export of fruits, vegetables and nuts, while they have a disadvantage in meats, cereals, and fish (except Morocco). Small farms are suited to producing labour-intensive crops, and the highest value per ha and per drop of water come from producing fruits, milk and vegetables.

Table 2.9. Coefficients of revealed comparative advantage for selected countries in the MENA region

	Egypt	Lebanon	Morocco	Jordan	Tunisia	Algeria
Vegetables	10.21	8.80	10.56	16.07		0.09
Fruits and nuts	6.71			4.53	3.36	0.09
Fish	0.15	0.06	3.00	0.08		
Meat	0.01	0.10	0.01		0.02	
Cereals		0.11	0.08		0.00	

Note: The table shows the Balassa Export Revealed Comparative Advantage Index (XRCA) applied to agricultural products. The XRCA is defined as the ratio of a product category's share in a country's total exports divided by the product category's share of global exports. An XRCA>1 implies that the country is specialised in the export of that product, while an XRCA<1 implies the opposite.
Source: Santos and Ceccacci (2015).

Food security situation

Households are food secure when they have year-round access to the amount and variety of safe foods their members need to lead active and healthy lives. Changes in food security, then, are driven mainly by events or conditions that affect families' ability to access safe food. Chief among these are incomes, the working of food markets to ensure food availability, and state public services to ensure food safety. The largest disrupter of these three factors in the region is conflict, which divides the region into two distinct sub-regions from the point of view of food security – conflict and non-conflict countries (Box 2.2).[13]

The Prevalence of Undernourishment (PoU) estimates the share of the population of a country facing absolute food deprivation. It is defined as the probability that a randomly selected individual from the reference population is found to consume less than his or her calorie requirement for an active and healthy life. (FAO, 2017c). Table 2.10 shows the prevalence of undernourishment in conflict and non-conflict countries in the MENA region.

As a rule of thumb, countries with a PoU of less than 5% are considered to be relatively food secure. As highlighted in Table 2.10, the non-conflict countries of the region are, in fact, relatively food secure. According to the PoU, in 2014-2016, the conflict countries of the MENA region were less food secure than the average level for least developed countries (LDCs). Whereas 28.2% of the population of the MENA conflict countries faced absolute food deprivation, only 24.4% of the population of the LDCs faced such insecurity (FAO, 2017c).

Though the high level of food insecurity in the conflict countries accords with expectations, care should be taken in interpreting these data for the prevalence of undernourishment. The PoU is a good indicator of hunger during periods when the income or consumption distribution is relatively constant, but it is not a good indicator of hunger when sharp changes in the distribution of food occur. The PoU likely underestimates the actual prevalence of undernourishment during times of conflict, because the inequality in food consumption parameters used to calculate it are derived from national household survey data, which are usually not available or accurate during times of conflict (FAO, 2017c).

Setting aside these caveats for the moment, the level of measured PoU in the conflict countries has been over three times the level in the rest of the MENA countries since 1999-2001, and has been rising gradually *vis-à-vis* the other countries in the region since 2003 (Table 2.10). This pattern in the evolution of the PoU in the conflict countries is consistent with it being partially driven by conflict, but it is also clear that they had relatively high levels of food insecurity even before conflict arose.

Table 2.10. Prevalence of undernourishment in conflict and non-conflict regions in MENA, 1999-2001 to 2014-16

	1999-2001	2001-2003	2003-2005	2005-2007	2007-2009	2009-2011	2011-2013	2013-2015	2014-2016
All MENA	9.7	9.8	10.0	10.0	9.6	8.9	8.4	8.4	8.8
Non-conflict countries	6.3	6.4	6.5	6.3	6.0	5.5	5.0	4.7	4.7
Conflict countries	29.0	28.4	28.9	29.1	28.5	26.6	25.3	26.1	28.2
Of which:									
--Yemen	29.9	30.7	30.9	28.9	27.1	25.7	24.6	25.2	28.8
--Iraq	28.3	26.6	27.4	29.3	29.6	27.2	25.9	26.7	27.8
--Sudan							25.9	25.7	25.6

Note: Undernourishment data exist for only three of the five conflict countries, and the aggregate is constructed from these data.
Source: FAO (2017c).

Box 2.2. Conflict and food security in the MENA Region

At the end of 2017, over 30 million people in this region were in need of assistance to satisfy their basic food needs. Among those, the food security situation was most critical in countries with lingering or escalating conflicts: Yemen, Syrian Arab Republic, Iraq and Sudan. In Yemen, according to the latest assessment carried out in March 2017, about 17 million people, corresponding to 60% of the total population, required food assistance. In the Syrian Arab Republic, some 6.5 million are estimated to be food insecure, and an additional 4 million at risk of food insecurity as they are using asset depletion strategies to meet their consumption needs. In Iraq and Sudan, about 3 million are food insecure. Smaller figures are reported for Libya and Mauritania, about 0.4 million each.

Residents in conflict zones often have to resort to food coping strategies to cover the severe food shortages they are facing. Households tend to reduce the number of meals and restrict the consumption of adults to prioritise children. If the crisis lingers, households deplete their assets and are no longer able to draw on stocks or other reserves. They resort to child labour, which often includes the withdrawal of children from school to carry out agricultural activities in order to cope.

Economic activity, including agricultural production, suffers in a conflict environment and further impairs livelihoods. While agricultural production is often one of the most resilient activities in an economy, those continuing to farm are often confronted with high production costs, lack of inputs and damaged or destroyed infrastructure. Agricultural activities, particularly those related to irrigated crops, suffer when fuel prices are high, with consequent increases in the share of rain fed crops, which in turn bear lower yields. Fertilisers are often subject to international sanctions. Farmers tend to plant seeds saved from the previous harvests, further constraining yields. Many rural households tend to rely on casual labour opportunities as their main source of income. In many conflict-affected areas, hired agricultural labour tends to be replaced by family labour in order to cope with the increased costs of production. While agricultural production improves household and local food availability, limited infrastructure including cold chain and transportation links often prevents deliveries to urban markets. Consequently, prices of local products tend to be low in producing regions, and high in the urban markets, despite availability.

The impact of lower agricultural production on world agricultural markets may be small, but has been dramatic in the affected countries. Before the conflict, Syria – one of the larger producers – produced on average about 4 Mt of wheat, but reached only 1.8 Mt in 2017. In Yemen, total domestic cereal production covers less than 20% of the total utilisation (food, feed and other uses). The country is largely dependent on imports from the international markets to satisfy its domestic consumption requirement for wheat, the main staple. The share of domestic wheat production in total food utilisation in the last ten years is between 5% to 10%, depending on the domestic harvest. While conflict did not substantially increase the country's dependence on imports, conflict-related decrease in production deteriorated livelihoods of farmers and pushed many to food insecurity.

The unpredictability of conflict threatens food security and local livelihoods but also livelihoods in the host countries. In addition to the millions who have fled countries due to the conflict, many are on the move internally, many multiple times. Internally displaced people and their host communities are often the most vulnerable to food insecurity. In Syria, about two in five people are on the move inside the country. In Iraq, in the first half of 2017, close to 1 million people were internally displaced, mostly due to the military operations in Mosul, in addition to the 3 million people already displaced by November 2016. As of early February 2018, over 5.5 million refugees were registered in the region covering Egypt, Iraq, Jordan, Lebanon and Turkey. In addition, a large share of the population lives abroad without seeking refugee registration.

Agricultural support policies

The vulnerability of countries to perceived risk from dependence on imported food has prompted some governments to support the cultivation of staple crops in the region (Box 2.3). Unfortunately, rigorous recent calculations of government support to (or implicit taxes on) producers have not been widely undertaken for the region, and to date have been made for only three countries with the latest year of available data from 2010. The nominal rate of assistance (NRA) is defined as the percentage by which government policies have raised gross returns to farmers above what they would be without the government's intervention (or lowered them, if NRA<0). The NRA considers only gross returns, and therefore does not consider input subsidies or taxes that may come through government-set prices for inputs. Estimates for wheat show a range of support from -28% (2010) in Sudan indicating effective taxation of their sector, to 44.7% in Egypt (2010) indicating strong support (World Bank, 2013). Support for wheat in Morocco was a more moderate 15% (2009). In addition to assistance to farmers, most countries in the region maintain consumer prices for selected types of bread and other staples at artificially low levels, effectively subsidising consumers. While these programs are often viewed as social support programs, they are extremely costly for government budgets and largely regressive (with the largest benefits accruing to the non-poor), and thus of dubious effectiveness and efficiency as social protection measures to reduce poverty. Between 2008 and 2013, the cost of non-targeted subsidies for fuel and food ranged from less than 1% in Lebanon to over 20% of GDP in the Islamic Republic of Iran. Though there have been efforts in most countries to reduce these subsidies since 2010, for most countries prices on energy products and basic foods are still controlled, albeit at a higher level, which reduces their fiscal impact (FAO, 2017c).

A comparison of farm gate producer prices and border import prices for wheat for the years since 2010 showed that producer prices in Algeria, Jordan, Kuwait, Oman, Saudi Arabia and Yemen were consistently significantly higher than the prices of imported wheat (from 60% to 250% higher). No firm conclusions can be based on these price differences, because the two prices are measured at different stages of the wheat value chain (producer prices at the farm gate and import prices at the border). However, such large price differences do suggest that domestic policies continue to raise prices for wheat above world prices.

Box 2.3. MENA government support for wheat

MENA governments have subsidised wheat production for many years using three main policy interventions: guaranteed prices, input subsidies, and import tariffs. The purpose of these policies is to raise the price and lower the costs for domestic production of wheat in order to increase self-sufficiency in wheat production.

In Iraq, for example, the Ministry of Trade supports wheat producers through a guaranteed price for no. 1 wheat that exceeds the import price of wheat. In 2015, the Ministry offered 795 000 Dinars (approximately, USD 681), in 2016, 700 000 Dinars (approximately USD 592) and in 2017, 560 000 Dinars (approximately USD 487) (USDA, 2017b). In Iran, the government also sets a minimum purchase price for wheat purchased by the state. State purchases at minimum prices have encouraged farmers to increase their production from 2.2 Mt in 2013 to 8.5 Mt in 2016. In Morocco in 2017, the government subsidised wheat production by establishing a reference price for purchasing domestic wheat (MAD 2 800 per tonne in 2017, equivalent to USD 286 per tonne). In October 2017, the government also introduced subsidies to millers and elevators that purchase domestic wheat. Furthermore, the government raised the import duty on soft wheat from 30% to 135% (Reuters, 2017). Tunisia's Cereal Board controls the marketing of 40% to 60% of total domestic wheat production and 10% to 40% of total barley production. The government sets guaranteed minimum prices for wheat and barley. For the 2017/18 marketing year, the Ministry of Agriculture set minimum prices of USD 329 per tonne for durum wheat and USD 236 for common wheat. The Ministry also subsidises irrigation water and provides technical advice to farmers targeted at increasing irrigated wheat area. Furthermore, in 2017, the Ministry subsidised agricultural machinery and irrigation equipment by 50% in order to encourage investment in irrigated cereals production (USDA, 2017a).

The Egyptian government heavily regulates wheat production, storage and marketing through many policy instruments. As of 2015, the Egyptian government subsidised the production of wheat through four main channels: (1) input and output subsidies for farmers, i.e. subsidised fertiliser prices and wheat procurement prices at higher than import prices; (2) consumer support in the form of highly subsidised prices for baladi bread; (3) government investment in improvements in grain storage and state grain trading; and (4) government support of wheat yield research, phytosanitary control, and other public goods. The government is also the sole purchaser of domestically produced wheat and imports about one third of total wheat imports. The government owns a large share of storage capacity and over half of the milling capacity of the country.

Saudi Arabia has undertaken the largest policy change. It gradually reduced its wheat production quotas and purchase programs because of strong concerns over the depletion of local water reserves which were used to irrigate wheat production. The country's production fell from around 2.5 Mt in 2005 to less than 30 000 t by 2015. Farmers have been encouraged to engage in alternative sustainable production activities such as greenhouse farming or production of fruits and vegetables using advanced drip irrigation techniques.

Sources: USDA (2017a, b); FAO and EBRD (2015); FAO (2017b); Reuters (2017).

Medium-term outlook

The previous sections introduced the food and agriculture and fish sectors of the MENA region and discussed the major issues the region has been facing. These include the region's challenges to improve food security and nutrition while sustainably raising productivity and managing the deepening dependence on foreign markets. This section expands on the discussion by exploring the potential future trends in consumption, production and trade of agriculture and fish commodities.[14]

Key economic and social factors shaping the outlook

The outlook for agriculture, food and fish in the MENA region is mainly driven by the region's macroeconomic performance, its demographic developments, the presence and extent of conflict and the evolution of policies.

According to World Bank data, on average, households in the region spend about 44% of their income on food and beverages.[15] Because of this high share, economic prospects will remain a critical factor affecting food consumption and food security in the coming decade. Based on the assumptions of improving energy markets, the continuation of structural policy reforms and no major changes in the favourable geopolitical climate, average income growth per capita in the region is projected at 1.6% p.a. for the coming decade, up from 1% p.a. in the previous one (Figure 2.11).[16] However, these income growth prospects are unlikely to lead to significant changes in dietary patterns.

Demographic developments are a second major driver of aggregate food demand. Population growth is expected to slow across the region, falling in aggregate from 2% p.a. in the last decade to 1.6% p.a. in the coming one (Figure 2.12), although this still represents almost 100 million additional people. The share of rural population is declining, but it will remain above 60% in the LDC countries while falling to around 10% in the Gulf region. The larger proportion of urban consumers will increase the demand for prepared foods, typically containing more fat and sugar.

Figure 2.11. Past and projected GDP per capita growth in the Middle East and North Africa

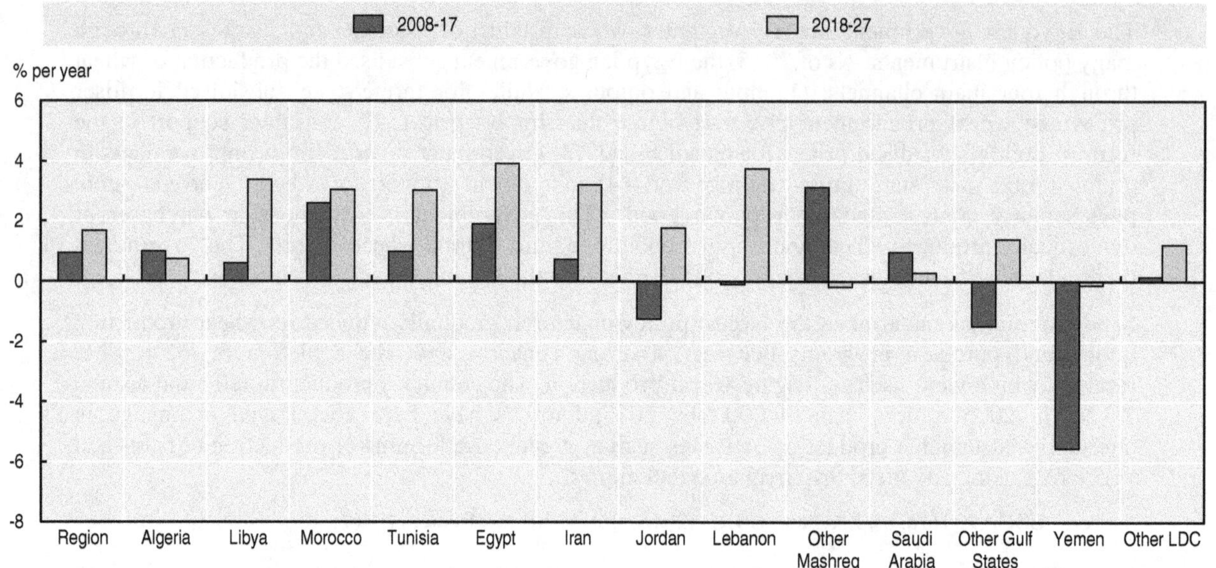

Source: OECD/FAO (2018), "OECD-FAO Agricultural Outlook", OECD Agriculture statistics (database), http://dx.doi.org/10.1787/agr-outl-data-en.

StatLink http://dx.doi.org/10.1787/888933742644

Figure 2.12. Population growth to slow, but unevenly across the region

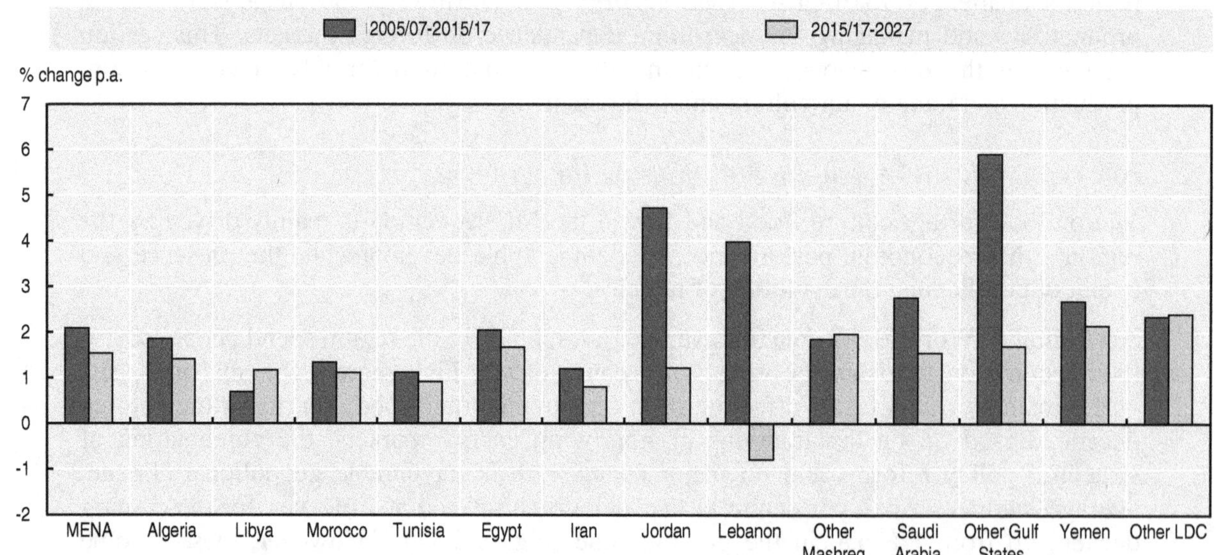

Source: World Population Prospects 2015: Revision from the UN Population Division and OECD/FAO (2018), "OECD-FAO Agricultural Outlook", OECD Agriculture statistics (database), http://dx.doi.org/10.1787/agr-outl-data-en.

StatLink http://dx.doi.org/10.1787/888933742663

Food consumption trends

Slow growth in per capita consumption

Food consumption in the region, measured in per capita calorie availability per day, is projected to grow at 0.4 % p.a., due mainly to modest income gains. Saturation effects in many high and medium income countries will slow consumption growth during the coming years, but higher growth (0.6% p.a.) is projected in the LDCs in the region, where it was stagnant or declining in the last decade. These improvements are predicated on higher income growth and no major changes in political stability. Average daily calorie availability (intake and consumer waste) per person in the region is projected to reach 3 200 kcal, varying from 3 440 kcal in the Gulf region, 3 412 kcal in North Africa, and 2 962 kcal in Other Western Asia to 2 420 kcal in the LDCs.

Diets in the MENA region are dominated by vegetal foods. The *Outlook* projects that animal foods will increase in share, due to higher meat, fish and dairy product consumption, but the transition will be slow (Figure 2.13). It is estimated that 89% of calories in the region will still stem from vegetal sources by 2027, only slightly down from the current level. Eating patterns across the region will remain relatively similar, and differences between the sub-regions are mainly due to their income differences. The countries of the Gulf region consume the highest share of animal foods at 15%. Second are the LDCs at 12%, as a result of their large animal husbandry sectors, while the countries of North Africa and Other Western Asia only reach about 10% by 2027. These shares of calories from animal sources compare with the stable 24% share which has been experienced in developed countries for many years.

Figure 2.13. Calories availability from various sources

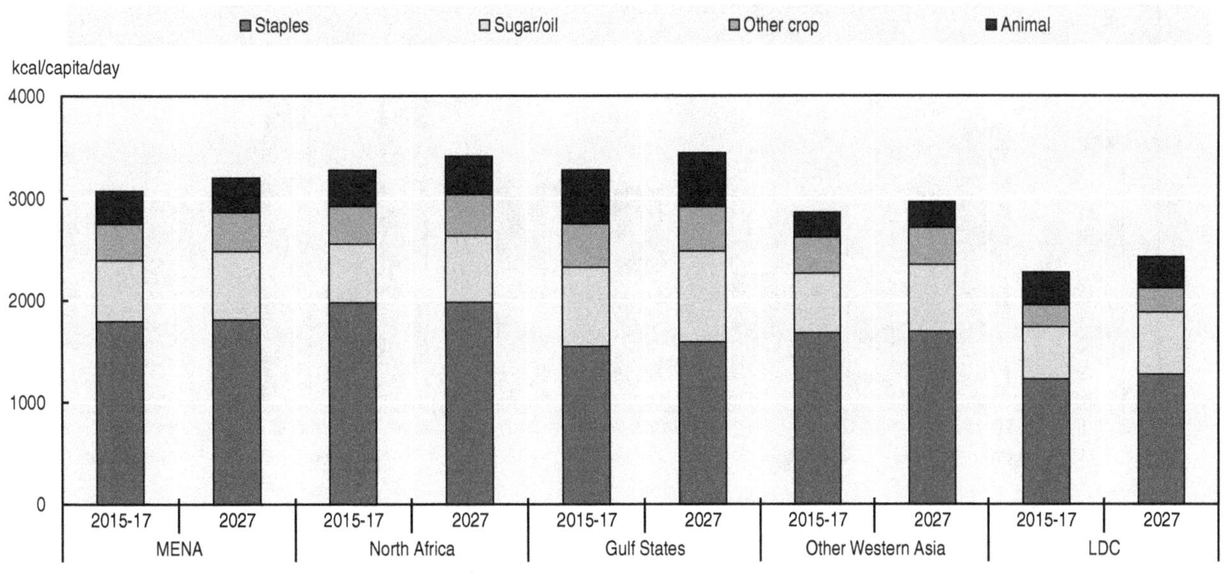

Source: OECD/FAO (2018), "OECD-FAO Agricultural Outlook", OECD Agriculture statistics (database), http://dx.doi.org/10.1787/agr-outl-data-en.

StatLink http://dx.doi.org/10.1787/888933742682

Dominance of cereals in diets will continue

Average annual food consumption of cereals is currently about 200 kg per person in the region, almost 60 kg higher than the world average. It is projected to stay roughly at this level over the projection period. Wheat is the traditional food staple in the region, yet its per capita consumption is projection to be flat. Rice is expected to show continued growth in the Gulf region, due to the consumption by migrants from southern and eastern Asia. In LDC countries, the use of locally grown coarse grains (primarily millet) is also expanding (Figure 2.14).

The share of calories from cereals in the diets continues to fall slowly, as growth in food demand comes from higher value products, especially vegetable oil and sugar.[17] The increased consumption of processed foods and prepared meals is expected to drive per capita vegetable oil use in the region from currently 19 kg to 22 kg per year by 2027. It will remain highest in the Other Western Asia region at 25 kg and lowest in the LDC countries where consumption will attain only 7 kg, as the population will still be largely rural and oilseeds are not grown locally.

Diets in the MENA region are traditionally very high in sugar and they are expected to stay that way, despite mounting health concerns. Consumption levels in countries such as Egypt, Saudi Arabia and Tunisia are around 40 kg/person/year. Average annual consumption of sugar is anticipated to grow, as lifestyles become more affluent, from 32 kg/person to 34 kg by 2027, at which level it will be on par with developed countries.

Figure 2.14. Wheat remains the most important cereal in the region

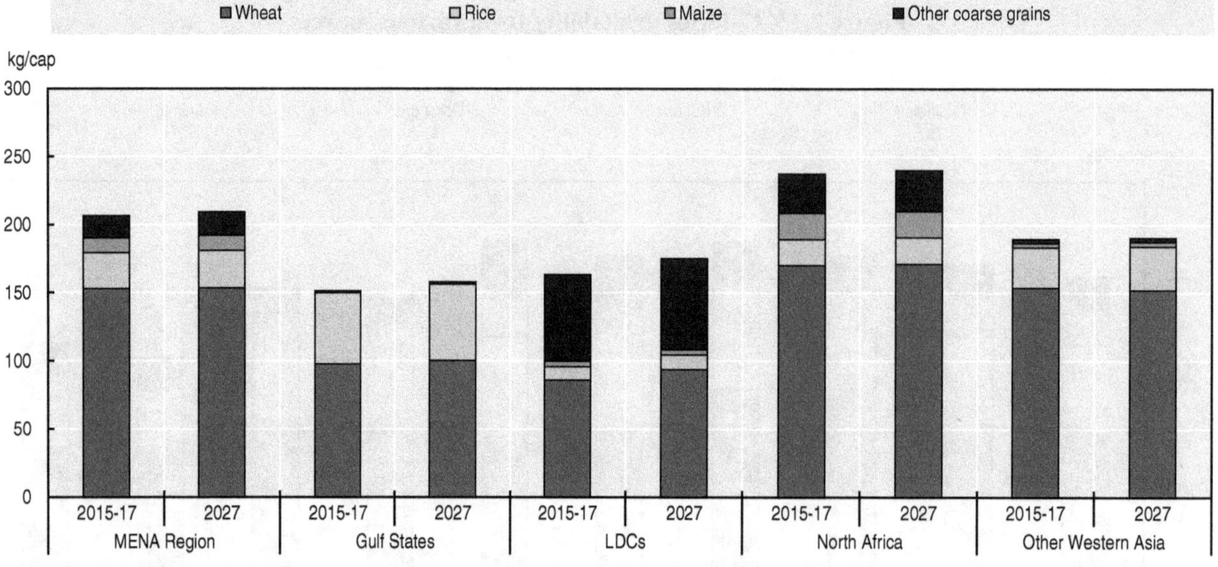

Source: OECD/FAO (2018), "OECD-FAO Agricultural Outlook", OECD Agriculture statistics (database), http://dx.doi.org/10.1787/agr-outl-data-en.

StatLink http://dx.doi.org/10.1787/888933742701

Low intake of proteins from animal sources

Meat is a distant second as a source of protein in the average MENA diet (Figure 2.15). The average meat consumption in the region is currently 25 kg/person p.a. (retail weight). Driven by income growth, it is projected to grow 0.6% per year over the medium term, led by growth in poultry, which is by far the most important meat consumed currently at 18 kg, growing at almost 1%. Meat consumption is highest in the Gulf region where it will increase marginally to 54 kg. Meat consumption in the LDC region will be largely driven by progress in the domestic sheep and cattle sector. It is expected to recover from recent declines to about 17 kg/person/year in 2027 based on projected productivity improvements by pastoralists.

Fish consumption in the MENA region has grown rapidly in recent years, at 4% p.a. in the last decade, and become second to poultry in providing protein in the MENA diet. While consumption is low and stagnant in LDC countries, growth elsewhere continues to outpace meat consumption.

Dairy products have become an important source of nutrition in the region, but per capita consumption fell in the last decade at a rate of 1.1% p.a. due to difficult production conditions especially in Other Western Asia and the Least Developed countries. In contrast, consumption grew strongly in the Gulf region at 4.9% p.a. and 1.8% in North Africa. Dairy product consumption continues to expand in the MENA region as producers enter more markets with a wider range of products. Fresh dairy products will continue to make up the largest share of the dairy market in the region, but there are growing markets for processed products including butter and cheese in more affluent countries. In lower income regions, particularly in North African countries, the demand for milk powders is significant. These are reconstituted into processed dairy products.

Figure 2.15. Share of animal protein in MENA diets is rising

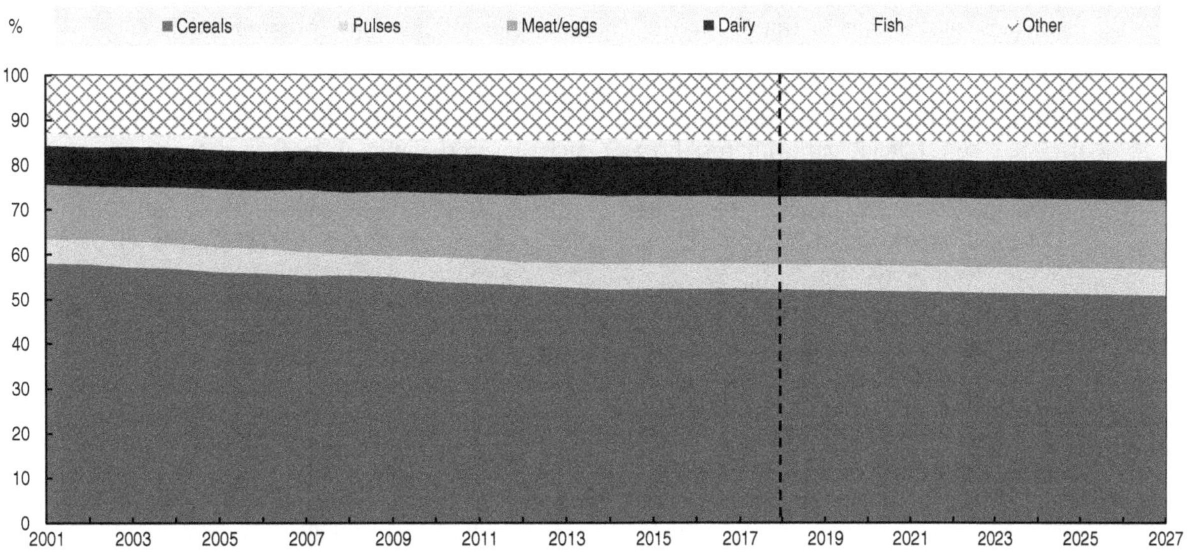

Source: FAOSTAT, OECD/FAO (2018), "OECD-FAO Agricultural Outlook", OECD Agriculture statistics (database), http://dx.doi.org/10.1787/agr-outl-data-en.

StatLink http://dx.doi.org/10.1787/888933742720

The outlook for production

The medium-term evolution of agricultural production in the MENA region will be shaped by a wide range of domestic and international factors. Agricultural production needs to address a series of domestic challenges in order to achieve sustainable development including aridity, limited cultivable land, scarce water resources and serious implications of climate change. Additionally, for almost all agriculture and fish products, price competition from international markets is high and in real terms, prices in these markets are trending down.

Due to these factors, agricultural and fish production in the region, measured in constant international prices, grew slowly at an annual rate of 1.3% p.a. over the last decade.[18] This slow rate of growth is due to falling real prices, but also to weak policies, insufficient investment in science and technology and agricultural development and conflict which have contributed to the impoverished state of agricultural resources and to their inefficient use and low productivity.

A modest improvement in production growth over the medium term is projected based on a generally improved economic setting, no deepening of conflict in some countries, and more stability in others which should improve investment and productivity. Average annual growth for the region as a whole is projected at 1.5% p.a. Critical to the region's growth prospects is the performance of its two main producer countries, Egypt and Iran, which together account for over half of the value of the MENA region's agricultural and fishery production. They are projected to grow 2.0% p.a. and 1.0% p.a. respectively.

Box 2.4. The future of food production in controlled environments

Many MENA countries are confronted with a dual challenge: they need to conserve their often small and fragile resource base, while also facing high and rising food import dependencies. Climate change will add to these challenges, further limiting production capacities and adding to import needs. These challenges are most pronounced in the countries of the Gulf Cooperation Council (GCC), where import dependencies can exceed 90% of domestic food needs and where both fertile cropland and renewable water resources are practically exhausted. In fact, many of these countries have grown food on irrigated desert land and with fossil water and, unsurprisingly, were forced to cease production completely soon after they had started. While adverse natural production environments have rendered these practices unsustainable, production in so-called "controlled environments" promises new and sustainable options to re-embark on domestic food production.

"Controlled environments" is a term commonly used to denote agricultural production independent of natural production environments. Typically, these are fully climate-controlled greenhouses, closed or semi-closed, where soil is replaced by an inert medium such as gravel or perlite and water supply is based on hydroponics. Nutrient supplies are managed either through fertiliser or "natural" plant nutrient sources such as animal or fish manure. Controlled environments are high-tech production plants that combine a whole range of different technologies, from fully automatic fertilisation, pest and weed control, robotic harvesting systems, LED lighting, solar-based heating, adiabatic cooling and energy-efficient desalination. They also use high ambient CO2 levels to boost yields, which can reach extra-ordinarily high levels of, for example, up to 100 kg of tomatoes/m^2. In analogy to smartphones, these production plants are also called "smart farms".

The combination of different technologies allows for location-independent and fully controlled production at high resource efficiency. These properties enabled controlled environments to make inroads in hot and arid environments, including the deserts of Arizona, Australia, and more recently also the GCCs.

The costs of production for some fruits and many vegetables are surprisingly low. Solar energy provides cheap electricity for cooling and LEDs, for desalination and N-fertiliser. CO_2 is available as a by-product from the hydro-carbon and cement industry, while migrant workers offer access to low cost labour for harvesting, grading and other labour-intensive processes. On the demand side, supermarkets provide cold chains and access to a large consumer base either through retail consumers or the large hospitality sector. Preliminary calculations suggest that products like tomatoes, eggplants, peppers or micro-herbs can be produced at costs of about 30-40% below the prices of airfreighted produce. A number of start-ups but also well-established companies are now seizing these new opportunities, apparent in the swiftly rising investments in controlled environments.

There are, however, risks and GCC-specific limits to producing in controlled environments. They include the need to have a highly-skilled operator ("head-grower") to run such a plant, the need to manage a complex supply chain from seedlings to spare parts, or the need to establish joint ventures with local partners, as foreign land ownership is heavily circumscribed or completely impossible in many GCCs.

Figure 2.16. Net value of agricultural production to grow more strongly

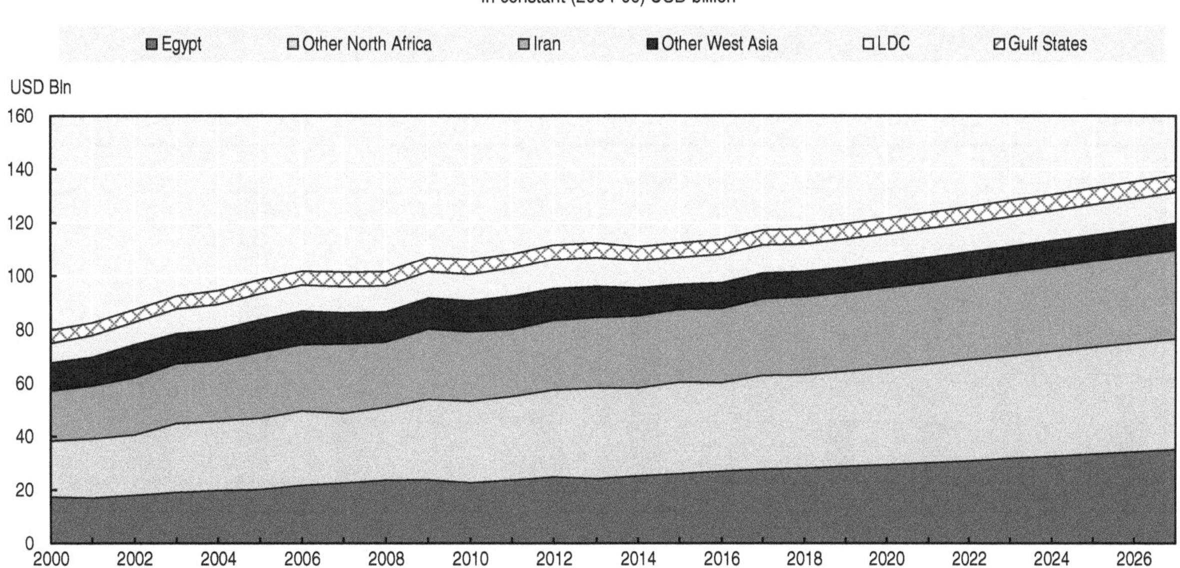

Source: FAOSTAT, OECD/FAO (2018), "OECD-FAO Agricultural Outlook", OECD Agriculture statistics (database), http://dx.doi.org/10.1787/agr-outl-data-en.

StatLink http://dx.doi.org/10.1787/888933742739

Agricultural production in the region is dominated by cereal production. While in the past production growth was achieved mainly by area expansion, yield improvements are seen as the major source of gains in the future. Cultivated land is projected to remain unchanged to 2027. Yields of the main crops, wheat, coarse grains and rice, are expected to grow at about 1.5% p.a., which is associated with improvements in seed potential, increased input intensity and improved management. Subsequently, the production of wheat, the region's major crop, is projected to reach 45 Mt by 2027, up from 37 Mt currently. Iran, the region's largest producer, will increase its share from 32% to 35% as its production will reach 16 Mt in 2027. Maize production, which fell in recent years due to severe declines in Iran, is set to recover in the medium term due to improved yields and will reach 10.5 Mt. Rice production, of which two-thirds takes place in Egypt, will attain 7.6 Mt by 2027, growing at 1.5% p.a., due to slower growth in cultivated area.

Sugar production, from sugar cane and increasingly sugar beets, has been the most rapidly growing commodity in the region. Sugar beet production grew rapidly at 6.4% p.a. in the last decade, underpinned by 10% p.a. area expansion in Egypt. It is projected to grow 3.0% p.a. over the outlook period, as sugar prices remain flat and fewer additional hectares will be cultivated. Growth in sugar cane production is mainly based on yield improvements, and is expected to grow slowly at about 0.8% p.a.

Milk production in the region was stagnant in the last decade, due to production declines in both Other West Asia and LDC countries, which were offset by growth in the other sub-regions. For the coming decade, the *Outlook* projects milk yield improvements of 1.6% p.a. and a cow herd enlargement by 0.2% p.a. As a result, milk production is expected to attain 38.4 Mt by 2027. Iran will continue to hold the largest share of production at about 20% followed by Egypt at 18%. As in the past, about 50% of milk will be consumed fresh, while 18% are going to be processed into cheese and 16% into butter, with the remaining share used in milk powder production.

Figure 2.17. Changes in major production activities in the MENA region

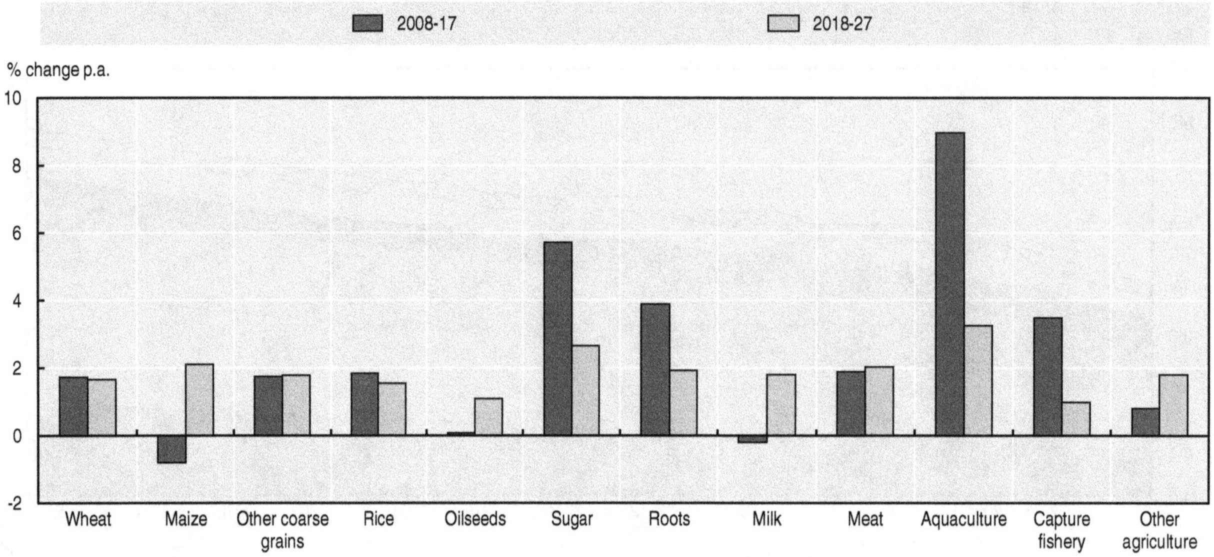

2008-17
2018-27
% change p.a.
10
8
6
4
2
0
-2
Wheat
Maize
Other coarse grains
Rice
Oilseeds
Sugar
Roots
Milk
Meat
Aquaculture
Capture fishery
Other agriculture

Source: OECD/FAO (2018), "OECD-FAO Agricultural Outlook", OECD Agriculture statistics (database), http://dx.doi.org/10.1787/agr-outl-data-en.

StatLink http://dx.doi.org/10.1787/888933742758

Current meat production in the region is about 10 Mt (carcass weight), with poultry meat accounting for about 60%, followed by bovine meat and sheep meat with about 20% each. Investments into new livestock production facilities together with higher carcass weights are expected to expand meat production across the region on average at 2.0% p.a., slightly higher than in the previous decade. In order to satisfy the fast growing domestic demand, poultry production is expected to increase at 2.8% p.a., led by strong growth in the North Africa region, where Egypt's poultry sector is dominant. The livestock sector of the LDC sub-region is characterised by a very large cattle inventory, which is currently estimated at about 45 million head, accounting for over 60% of the total MENA cattle inventory. Nevertheless, due to traditional herding practices with low offtake ratios, the sub-region produces just 22% of the bovine meat of the MENA region.

Capture fisheries still dominates fish production in the MENA region. Currently, almost 4 Mt are landed per year, with Morocco accounting for almost 40%. Growth in the next decade is limited to 0.5% p.a. due to dwindling fish stocks. Aquaculture production in the region more than doubled during the last decade, standing currently at almost 2 Mt. It is set to increase by another 50% over the ten years, with growth expected in all sub-regions, particularly North Africa (Egypt) which contributes 75% of the total supply.

The outlook for trade

The MENA region is one of the largest net food importing regions of the world, with significant net imports of virtually all food commodities; trade has been and will remain the most important contributor to additional food supplies in the region. Currently, about 27% of international shipments of cereals, 21% of sugar, 20% of poultry meat, 39% of sheep meat, 20% of skim milk powder and 30% of whole milk powder go to the MENA region. Domestic markets in the region are generally tightly integrated into global agricultural markets and this interdependence is certain to continue and expected to deepen for products such as wheat and maize.

Large increases in net imports are projected, as consumption will continue to outpace production for most basic food commodities. The deficit is projected to reach 58 Mt for wheat and 65 Mt for coarse grains in 2027. The largest share of MENA imports for almost all commodities will continue to go to North Africa, followed by Other Western Asia. Other coarse grains and rice are the exception, as the Gulf region dominates for them (Figure 2.19). The Gulf region dominates meat and fish imports, given its low production and relatively high consumption levels. The LDC countries are the only net exporter of fish in the region, and these exports are projected to increase.

Figure 2.18. Rising net imports for all commodities and in all regions

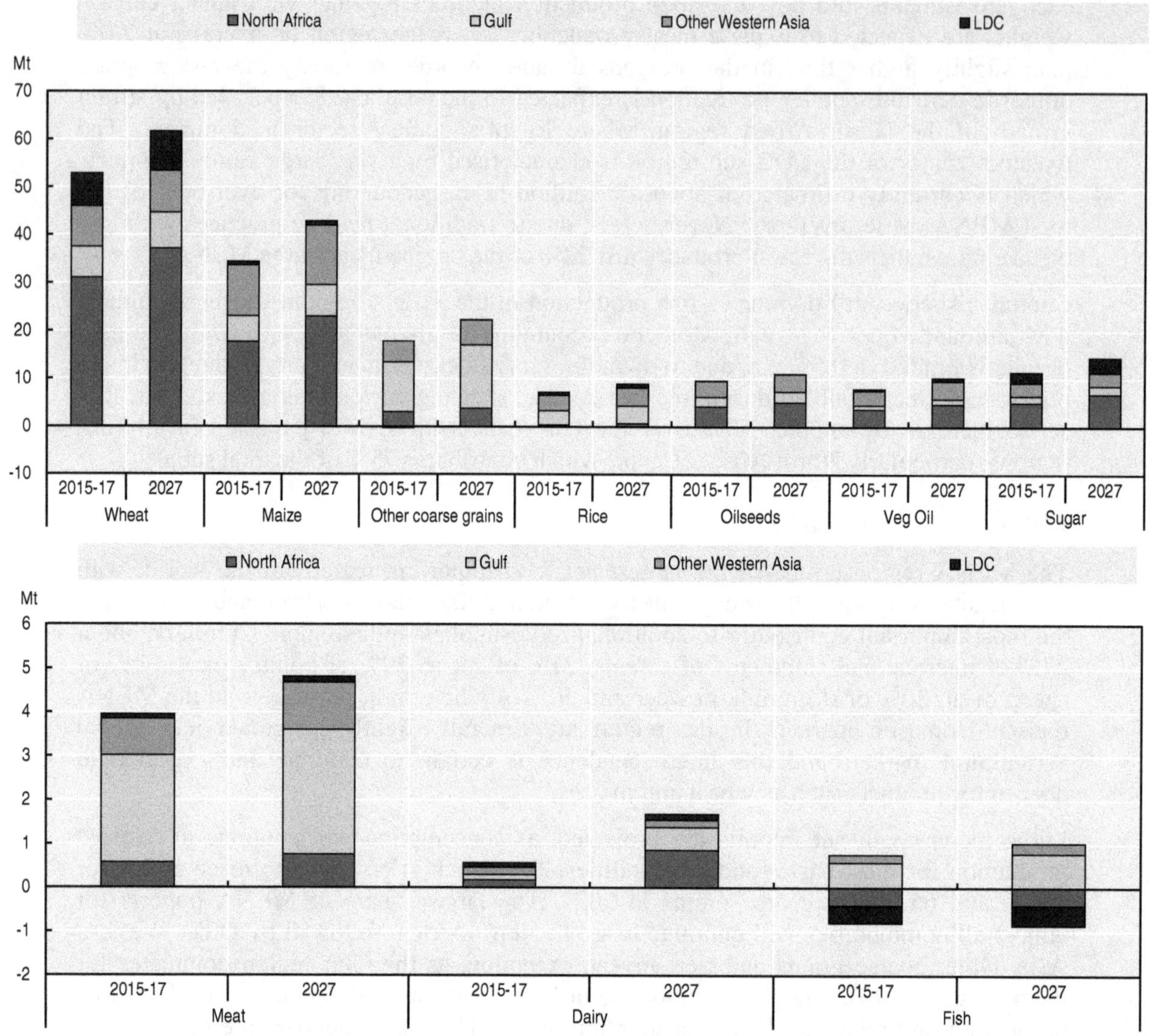

Source: OECD/FAO (2018), "OECD-FAO Agricultural Outlook", OECD Agriculture statistics (database), http://dx.doi.org/10.1787/agr-outl-data-en.

StatLink http://dx.doi.org/10.1787/888933742777

Figure 2.19. High dependence on foreign markets for basic foodstuffs

Self sufficiency ratio, production/consumption

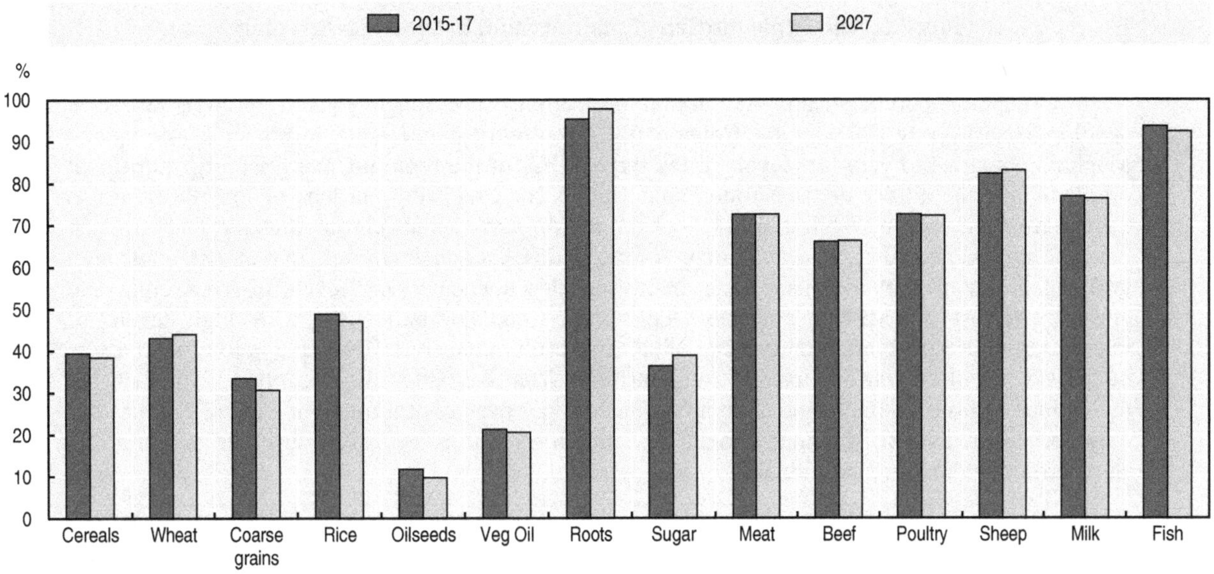

Source: OECD/FAO (2018), "OECD-FAO Agricultural Outlook", OECD Agriculture statistics (database), http://dx.doi.org/10.1787/agr-outl-data-en.

StatLink http://dx.doi.org/10.1787/888933742796

Risks and uncertainties

The medium-term outlook projections for the Middle East and North Africa region are subject to risks and uncertainties associated with internal and external issues. Conflicts critically impact food consumption as well as agricultural production. Other uncertainties include, for example, nutritional concerns or volatility in crude oil prices. These issues are analysed below to illustrate their potential impact on the projections.

Addressing nutritional concerns

Parts of the MENA region face what is referred to as the "triple burden" of malnutrition: undernourishment, over-nourishment or obesity, and malnutrition (Box 2.5). Albeit slowly, undernourishment is diminishing, at least where conflicts are not present. But the latter two nutritional outcomes are rising, and governments are considering policy measures to address these.

The United Nations report "Arab Horizon 2030" undertook scenario analysis to examine a radical change in diets of the Arab region (which broadly corresponds to the MENA region as defined here, but excludes Iran).[19] Addressing the diet problem has implications for the dependence on foreign markets for basic foodstuffs. A so-called "Healthy Diet Scenario" was constructed that assessed the impacts of an improved diet on domestic and international markets. Using the OECD-FAO Aglink–Cosimo model, a scenario was simulated in which eating patterns were assumed to conform to FAO and WHO "healthy diet" recommendations of 2 200 kilocalories per day which would be achieved through a 50% decrease in cereal availability for food consumption, a doubling of meat and egg consumption, a tripling of dairy products and reduced sugar and vegetable oil consumption. Assuming a "waste" factor of 30% that is implicit in the baseline calorie availability estimate, such changes involve a decrease in total caloric availability from 3 100 kcal per day to 2 860 kcal per day.

Box 2.5. The triple burden of malnutrition in the MENA region

The Middle East and North Africa (MENA) region comprises 22 countries at very different levels of development, income, health and social protection.[1] The range goes from high levels of development in Gulf Cooperation Council (GCC) countries, moderate levels in Mashreq and Maghreb countries to very low levels in the three LDCs of the region. Unsurprisingly, nutritional problems and the ability of the various countries to cope with the burdens of malnutrition also differ across the region. While the LDCs of the region face serious chronic hunger challenges or outright famines, the GCC alongside many of the middle income countries, in contrast, confront a growing problem of over-consumption and, as a consequence, rising levels of overweight and obesity. Almost all MENA countries have rather undiversified diets with high levels of micronutrient deficiencies, notably iron, which can result in anaemia. The table below summarises the prevalence of the various forms of malnutrition. What the table fails to capture is the fact that the various forms of malnutrition are not confined to, nor even concentrated in a given country, but they occur simultaneously in many countries, sometimes within the same household and in a few cases afflicting the same individual

	Middle East		North Africa*	
	2005	2015	2005	2015
			%	
Prevalence of undernourishment in the total population	9.1	9.1	4.6	6.7
Prevalence of food insecurity in the adult population (>=15)	30.9	8.7	27.9	11.2
Prevalence of wasting in children (< 5y)		3.9		7.9
Prevalence of stunting in children (< 5y)	20.6	15.7	21.6	17.6
Prevalence of overweight in children (< 5 y)	7.0	8.0	8.9	10.0
Prevalence of obesity in the adult population (>=18 y)	20.3	25.8	17.5	22.6
Prevalence of anaemia among women of reproductive age (15-49)	34.1	37.6	36.7	32.6

* Including Sudan.

The simultaneous occurrence of the various forms of malnutrition is known as the "triple burden of malnutrition." It takes a growing toll on the region's health sector and even on overall economic performance. On the one hand, anaemia and undernourishment reduce a person's ability to undertake physical work and thus can create poverty traps, particularly but not only in LDCs. On the other hand, overweight and obesity have become increasingly visible through the high prevalence levels of non-communicable diseases (NCDs), observed notably in GCC but also in Mashreq and Maghreb countries.

The simultaneous occurrence of the various forms of malnutrition also makes it difficult to address the three problems efficiently. Past programmes often took a "wholesale" approach, e.g. by lowering food prices for all consumers, particularly for basic foods (bread/flour/sugar). While this resulted in improved access to basic food energy even for the poorest consumers, it also added to a growing problem of overweight and obesity and, not unrelated, food waste. A number of factors make policy choices particularly difficult for the MENA region. They include high inequality of wealth and incomes, hence different responsiveness to price and policy incentives; high shares of migrant populations and different ethnicities particularly in the GCC, hence different genotypic predispositions to develop NCDs; weak institutions as well as deficiencies in food delivery systems and physical infrastructure, hence rendering the administration of food supplementation and fortification programmes difficult. Consequently, addressing the triple burden demands much more targeted and innovative policy instruments than those applied in the past.

1. Estimates are adjusted to definitions of indicators and regions based on data from the *State of Food Insecurity and Nutrition* (FAO, 2017f).

The effect on domestic production was simulated under the assumption of an unconstrained expansion of the supply in the region. Under this scenario, meat production in the Arab region would, by 2030, increase from 2 Mt to 13 Mt in the healthy diet scenario. Dairy product production (fluid milk equivalent) would increase from 5 Mt to 25 Mt by 2030. Though food consumption of cereals under the healthy diet scenario would decrease substantially, overall demand for cereals under the scenario would increase. The large increase of the livestock sector and subsequent domestic feed use of grains would drive this increase. Feed demand for cereals would grow six times faster under the healthy diet scenario than under the business as usual baseline. Production of feed within the Arab region would not be able to grow this fast, so that the region would require additional feed imports. Consequently, the self-sufficiency rate for cereals would be lower under the healthy diet scenario than under the baseline projection.

While such a substantial change in the average diet would affect the nutrition status of the average consumer in the Arab region in a positive way, it would not lessen the region's dependence on foreign markets, as either feed grains or alternatively livestock products would have to be imported.

Analysis of alternative crude oil price projections

The foreign exchange balance of many MENA countries is critically influenced by crude oil prices. A simulation as presented in the Overview, using a rise in the crude oil price to USD 122/barrel rather than the baseline value of USD 76/barrel by 2027, illustrates the significance of oil prices for the region. Figure 2.20 illustrates the estimated impacts on consumption and trade. Higher oil prices lead to higher world reference prices for cereals of around 10%, which in turn lead to higher retail prices in MENA of about 6%. The estimated increases in per capita GDP range from 2% for Egypt to 15% for Saudi Arabia. The result is that on average for the region, daily calorie availability increases by 0.6% in 2027, meaning that the income effect of higher oil prices outweighs the hike in food prices, generating overall higher food consumption in the region. Among the least developed countries of the region, Yemen's estimated increase in GDP is 8% by 2027, which leads to a 2.5% increase in calorie intake. The estimated trade impacts for cereals vary by country but for the region as a whole, net wheat imports increase marginally.

Figure 2.20. Impact of higher oil prices on food prices consumption and trade

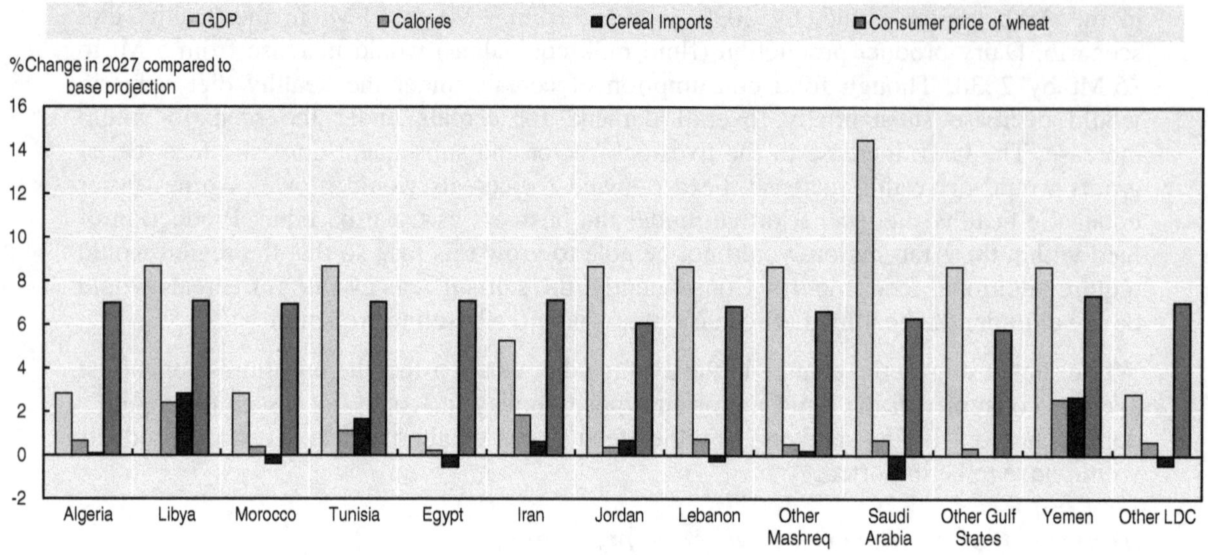

Source: OECD/FAO Secretariats.

StatLink http://dx.doi.org/10.1787/888933742815

Implications for food security prospects in the region

According to recent estimates (FAO, 2017f) for 2014-16, the prevalence of undernourishment is highest in the region for Sudan (25.6%), Iraq (27.8%), and Yemen (28.8%), with no reliable data for Syria. Projections of higher calorie and protein availability, based on the assumption of stable economic development and a stable income distribution, should imply a decline of the prevalence of undernourishment over time, particularly in the least developing countries.

Conclusions

The outlook for the MENA region assumes little change in agricultural, natural resource and economic growth policies. Its implications for the region are that food demand, supply and trade outcomes will continue along a similar trajectory that has been observed in the past—slow growth in food consumption, gradual changes in diet to include higher consumption of animal products, continued water use at unsustainable rates, and continued and increasing reliance on world markets. The main difference to past trends would be higher meat, milk, maize and oilseeds production associated with higher consumption of animal proteins. While increasing maize and milk production represent a recovery from quite poor performance over the past decade, increasing meat production is based on the assumption that an improved economic environment will lead to more investments and subsequent productivity improvements in the region. These developments are anticipated to limit, but not reverse, increases in the dependence of the region on imports.

Current agricultural policies in the region emphasize wheat price support bolstered by import protection (Box 2.1). These policies are aimed at limiting import dependency for cereals. At the same time, consumer policies emphasise subsidised prices for staple foods and are viewed as social protection measures. The results of these policies can be seen in the pattern of harvested land, of which 60% remains in water-thirsty cereals.

An alternative approach to food security and agricultural policies would emphasise rural development, support for production of higher-value horticulture products on small farms, supported by a more robust technical extension system. This approach is rooted in the conviction that the level of food security of a country hinges more on the elimination of poverty than on wheat self-sufficiency. Fruits and vegetables both consume less water and provide higher economic returns per drop, and many countries in the region have a comparative advantage in their production. While such higher-value crops and livestock products could potentially increase farmer incomes, improve nutrition, and use water more sparingly, they require a higher level of agronomic and export market knowledge and present higher levels of risk. A revision of food security policies away from self-sufficiency towards poverty elimination would focus the attention of policymakers on rural development and on building the capacity of farmers to minimise risk while raising higher value crops.

From a nutritional perspective, diets in the MENA region will remain very rich in in cereals, and wheat in particular. The share of vegetable oil and sugar, as well as meat, fish and dairy products will grow, albeit slowly. Barring increased conflict, undernourishment should decline slowly as average food consumption levels increase. However, the evolution of diets is also expected to contribute to increased rates of obesity with associated health consequences. The current structure of policy support toward consumers of cereals limits needed diet diversification, and should be altered to redress rising health issues.

Notes

[1] In this chapter, the Middle East and North Africa region includes countries/areas of FAO's North Africa and Near East region: Algeria, Bahrain, Egypt, Iran, Iraq, Jordan, Kuwait, Lebanon, Libya, Mauritania, Morocco, Oman, the Palestinian Authority, Qatar, Saudi Arabia, Sudan, Syria, Tunisia, United Arab Emirates (UAE), and Yemen

[2] Water stress is indicated when annual freshwater withdrawals are high compared to renewable internal freshwater resources. If freshwater withdrawals exceed renewable internal resources then either non-renewable groundwater resources are being withdrawn or desalinated and other supplemental water resources are being used that are not included in the total annual water resources figures (World Bank, 2018).

[3] The Herfindahl-Hirschmann concentration index, is a measure of the degree of product concentration. The index ranges between 0 and 1. An index value closer to 1 indicates a country's exports or imports are highly concentrated on a few products. On the contrary, values closer to 0 reflect exports or imports are more homogeneously distributed among a series of products. For worldwide evidence of systematically high concentration index values for natural resource rich countries, see Bahar (2016).

[4] Sodicity refers to high concentrations of sodium in soils. Sodic soils have a poor structure as sodium causes soils to swell and disperse. A dispersed soil structure loses its integrity, becomes prone to waterlogging, and is usually harder, making it difficult for roots to penetrate.

[5] The value of gross production includes all livestock and crop production, including crops used for feed. The proper land comparator for gross production is agricultural land, which includes both arable land and pastures.

[6] All values are expressed in dollars using average international prices of 2004-2006.

[7] The "water scarcity line" is defined in UNDP (2006).

[8] Generally, water "used" means that it (1) is depleted through evapotranspiration; (2) is absorbed into a product; (3) flows to a location where it cannot be readily reused; or (4) becomes heavily polluted (Molden et al., 2010).

[9] Cline (2007). Calculations are based on the IPCC Third Assessment Report published in 2001.

[10] Verner and Breisinger (2013); FAO (2015); Ward and Rucksthuhl (2017).

[11] Cereal area is mostly planted with wheat. In 2014, of total cereal area, wheat accounted for 43%, sorghum 23%, barley 18% and millet 8%. The current mix of wheat and coarse grains area is only slightly different from that in the 1960s when wheat made up half of all harvested area of cereals.

[12] The Gulf Cooperation Council includes the countries of Bahrain, Kuwait, Oman, Qatar, Saudi Arabia and the United Arab Emirates.

[13] Conflict countries include Sudan, Syria, Yemen, Libya and Iraq.

[14] In this section, countries are often aggregated into regional groups. The North Africa region is Morocco, Algeria, Tunisia, Libya and Egypt. The Gulf region includes the states of the Gulf Cooperation Council: Bahrain, Kuwait, Oman, Qatar, Saudi Arabia, and UAE. The Other Western Asia region includes Iran, Lebanon, Jordan and other Mashreq countries of Syria, the Palestinian Authority, and Iraq. The least developed country (LDC) region includes Yemen, Sudan and Mauritania.

[15] See www.worldbank.org, Global Consumption Database. Shares are based on 2016 values.

[16] See IMF *World Economic Outlook*, January, 2018, and World Bank Global Economic Prospects, January, 2018 for more detailed discussion.

[17] Excluding olive oil which is not included in this projection.

[18] See FAOSTAT "Net agricultural production" which weights agricultural production of each commodity by international reference prices during the period 2004-06. The value of production is net of the value of seed and feed inputs. The value of fish production is added, and is need of feed inputs.

[19] Based on Food and Agriculture Organization of the UN (FAO) and United Nations Economic and Social Commission on West Asia (ESCWA). 2018. Arab Horizon 2030 (Beirut, ESCWA).

References

Bahar, D. (2016), "Diversification or Specialization: What is the Path to Growth and Development?" *Global Economic and Development at Brookings Policy Brief* (November) (https://www.brookings.edu/wp-content/uploads/2016/11/global-20161104-diversification.pdf).

Bucchignani, E., et al.(2018), "Climate Change Projections for the Middle East - North Africa Domain with COSMO-CLM at Different Spatial Resolutions," *Advances in Climate Change Research,* available online 9 February 2018 (https://ac.els-cdn.com/S1674927817300552/1-s2.0-S1674927817300552-main.pdf?_tid=ed36ecde-2036-48e9-95ce-6d94afee3190&acdnat=1520339775_7fe177ee762aaafd47500e3a5b6dbe62).

Cline, W. (2007), *Global Warming and Agriculture: Impact Estimates by Country*, Peterson Institute, Washington, DC, Peterson Institute.

Elhadary, Y. and H. Abdelatti (2016), "The Implication of Land Grabbing on Pastoral Economy in Sudan" *World Environment 2016*, Vol. 6(2), pp. 25-33 http://article.sapub.org/10.5923.j.env.20160602.01.html.

Food and Agriculture Organization of the UN (FAO) and United Nations Economic and Social Commission on West Asia (ESCWA) (2018a), *Arab Horizon 2030*, ESCWA Publications, Beirut.

Food and Agriculture Organization of the UN (FAO) (2018b), Aquastat Main Database, http://www.fao.org/nr/water/aquastat/data/query/index.html?lang=en.

Food and Agriculture Organization of the UN (FAO) (2018c), FAOSTAT Database http://www.fao.org/faostat/en/#data.

Food and Agriculture Organization of the UN (FAO) (2018d), *Global Agro-Ecological Zones (GAEZ)*, FAO Publications, Rome, http://gaez.fao.org/Main.html#.

Food and Agriculture Organization of the UN (FAO) (2017a). *Morocco: GIEWS Country Brief,* FAO Publications, Rome, http://www.fao.org/giews/countrybrief/country.jsp?code=MAR.

Food and Agriculture Organization of the UN (FAO) (2017b), *Middle East and North Africa Regional Overview of Food Security and Nutrition, Building Resilience for Food Security and Nutrition in Times of Conflict and Crisis*, FAO Publications, Cairo, http://www.fao.org/3/I8336EN/i8336en.pdf.

Food and Agriculture Organization of the UN (FAO) (2017c), "Regional Review on Status and Trends in Aquaculture Development in the Near East and North Africa – 2015, *FAO Fisheries and Aquaculture Circular*, No. 1135/6, FAO Publications, Rome.

Food and Agriculture Organization of the UN (FAO) (2017d), "Fisheries and Aquaculture: Implementation of the Blue Growth Initiative in the Near East and North Africa", Doc 16/5, Thirty-third Session of the FAO Regional Conference for the Near East, 2016.

Food and Agriculture Organization of the UN (FAO) (2017e), *State of Food Insecurity and Nutrition in the World 2017*, FAO Publications, Rome.

Food and Agriculture Organization of the UN (FAO) (2015a), "Food Security and Sustainable Agriculture in the Arab Region" Regional Coordination Mechanism (RCM) Issues Brief for the Arab Sustainable Development Report, FAO Publications, Rome, http://css.escwa.org.lb/SDPD/3572/Goal2.pdf.

Food and Agriculture Organization of the UN (FAO) and European Bank for Reconstruction (2015b) "Focusing on Comparative Advantage" in *The Agrifood Sector in the Southern and Eastern*

Mediterranean: A Collection of Notes on Key Trends, FAO Publications, Rome, http://www.medagri.org/docs/group/19/Agribusiness%20Notes_web.pdf.

Food and Agriculture Organization of the UN (FAO) and European Bank for Reconstruction and Development (EBRD) (2015), *Egypt Wheat Sector Review*, FAO Publications, Rome,http://www.fao.org/3/a-i4898e.pdf.

Fuglie, K. and N. Rada (2013), "Growth in Global Agricultural Productivity: An Update" *Amber Waves*, 18 November, https://www.ers.usda.gov/amber-waves/2013/november/growth-in-global-agricultural-productivity-an-update/.

Hulton, C. (2000), "Total Factor Productivity: A Short Biography," *NBER Working Paper* No. 7471, http://www.nber.org/papers/w7471.pdf.

Jones, A., et al (eds.) (2013). Soil Atlas of Africa, European Commission Publications, Luxembourg, https://esdac.jrc.ec.europa.eu/content/soil-map-soil-atlas-africa.

Jouili, M. (2009), "Tunisian agriculture: Are small farms doomed to disappear?" Paper presented at the 111 EAAE-IAAE Seminar 'Small Farms: decline or persistence,' University of Kent, Canterbury, 26-27 June, https://hal.archives-ouvertes.fr/hal-01180353/document.

Lowder, S., J. Skoet, and T. Raney (2016), "The Number, Size, and Distribution of Farms, Smallholder Farms, and Family Farms Worldwide" *World Development*, Vol. 87, https://www.sciencedirect.com/science/article/pii/S0305750X15002703.

Mekonnen, M. and A. Hoekstra (2010b), *The Green, Blue and Grey Water Footprint of Farm Animals and Animal Products, Vol. I: Main Report*, Value of Water Research Report Series No. 48, UNESCO-IHE Institute for Water Education, Delft, the Netherlands, https://ris.utwente.nl/ws/portalfiles/portal/6453582.

Mekonnen, M. and A. Hoekstra (2010a), *The Green, Blue and Grey Water Footprint of Crops and Derived Crop Products, Vol. 2: Appendices*, Value of Water Research Report Series No. 47, UNESCO-IHE Institute for Water Education, Delft, the Netherlands, https://ris.utwente.nl/ws/portalfiles/portal/6453584.

Molden, D., et al. (2010), "Improving Agricultural Water Productivity: Between Optimism and Caution" *Agricultural Water Management*, Vol. 97, http://www.icarda.org/wli/pdfs/articles/4-ImprovingAgriculturalWaterProductivity.pdf.

Nuno Santos, N. and I. Ceccacci (2015), *Egypt, Jordan, Morocco and Tunisia: Key Trends in the Agrifood Sector*, FAO Publications, Rome, http://www.fao.org/3/a-i4897e.pdf.

Rae, J. (no publication date), *An Overview of Land Tenure in the Middle East Region*, FAO Publications, Rome, http://www.fao.org/3/a-aq202e.pdf).

Reuters (2017). "Morocco Introduces Measures to Support Wheat Output", 9 May, https://www.agriculture.com/markets/newswire/morocco-introduces-measures-to-support-local-wheat-output-statement.

Rucksthuhl, S. and C. Ward (2017), *Water Scarcity, Climate Change and Conflict in the Middle East: Securing Livelihoods, Building Peace*, I.B. Tauris and Company, London.

Sdralevich C. et al. (2014), *Subsidy Reform in the Middle East and North Africa Recent Progress and Challenges Ahead*, International Monetary Fund, Washington, D.C., https://www.imf.org/external/pubs/ft/dp/2014/1403mcd.pdf.

UN Statistics Division (UNSD) (2018), UN National Accounts Main Aggregates Database (https://unstats.un.org/unsd/snaama/Introduction.asp).

UNCTAD (2018), UNCTAD Stat, http://unctadstat.unctad.org/wds/ReportFolders/reportFolders.aspx?sCS_ChosenLang=en.

United Nations Development Programme (UNDP) (2006), *UN Human Development Report 2006: Beyond scarcity: Power, Poverty and the Global Water Crisis*, UNDP, New York, http://hdr.undp.org/sites/default/files/reports/267/hdr06-complete.pdf.

US Department of Agriculture (USDA). 2017a. "A Strong but Fatigued 2017 Campaign," *Global Agricultural Information Network (GAIN) Report: Tunisia Grain and Feed Annual* (4/11/2017), https://gain.fas.usda.gov/Recent%20GAIN%20Publications/Grain%20and%20Feed%20Annual_Tunis_Tunisia_4-11-2017.pdf.

US Department of Agriculture (USDA) (2017b), "Iraqi Wheat Production Down; Weather, Procurement Drop, and Conflict to Blame," *Global Agricultural Information Network (GAIN) Report: Iraq Grain and Feed Annual* (10/10/2017) https://gain.fas.usda.gov/Recent%20GAIN%20Publications/Grain%20and%20Feed%20Annual_Baghdad_Iraq_10-10-2017.pdf.

Verner, D. and C. Breisinger (2013) *Economics of climate change in the Arab world: case studies from the Syrian Arab Republic, Tunisia and the Republic of Yemen*, World Bank, Washington, D.C. https://openknowledge.worldbank.org/bitstream/handle/10986/13124/763680PUB0EPI0001300PUBDATE03021013.pdf?sequence=1&isAllowed=y.

World Bank (2013), "Updated National and Global Estimates of Distortions to Agricultural Incentives, 1955 to 2011", World Bank, Washington, D.C. http://siteresources.worldbank.org/INTRES/Resources/469232-1107449512766/UpdatedDistortions_to_AgriculturalIncentives_database_0613.zip.

World Bank (2018), *World Development Indicators*, World Bank, Washington, D.C. http://databank.worldbank.org/data/reports.aspx?source=world-development-indicators.

World Bank (2018), *Beyond Scarcity: Water Security in the Middle East and North Africa*, MENA Development Report, World Bank, Washington, D.C. https://openknowledge.worldbank.org/handle/10986/27659.

OECD PUBLISHING, 2, rue André-Pascal, 75775 PARIS CEDEX 16
(51 2018 04 1 P) ISBN 978-92-64-29721-0 – 2018

www.ingramcontent.com/pod-product-compliance
Lightning Source LLC
LaVergne TN
LVHW081413110826
845149LV00010B/1732

* 9 7 8 9 2 6 4 2 9 7 2 1 0 *